Vaxxers

'This is one of the most epic and pioneering moments in human history, comparable to the race to put a man on the moon, the discovery of DNA, or the first ascent of Everest. The Oxford AstraZeneca vaccine is a triumph and its creators are life savers. Science is the exit strategy, as long as we make that science equitably available to the world – as all the incredible people behind the Oxford AstraZeneca vaccine always intended. Truly the "People's Vaccine".'

Sir Jeremy Farrar, Director of the Wellcome Trust

'What an enthralling tale of toil, tenacity and triumph this is. The authors' intelligence, idealism and sheer, bloody-minded grit shine through. The world needs all the Sarah Gilberts and Catherine Greens it can get. Just brilliant.'

Dr Rachel Clarke, author of *Dear Life* and *Breathtaking*

Vaxxers

The Inside Story of the Oxford AstraZeneca Vaccine and the Race Against the Virus

PROFESSOR SARAH GILBERT

AND

DR CATHERINE GREEN

Written with Deborah Crewe

First published in Great Britain in 2021 by Hodder & Stoughton
An Hachette UK company

7

A CIP catalogue record for this title is available from the British Library

Hardback ISBN 9781529369854
Trade Paperback ISBN 9781529369878

Typeset in Bembo by Palimpsest Book Production Ltd, Falkirk, Stirlingshire

Printed and bound in Great Britain by Clays Ltd, Elcograf S.p.A.

Hodder & Stoughton policy is to use papers that are natural, renewable and recyclable products and made from wood grown in sustainable forests. The logging and manufacturing processes are expected to conform to the environmental regulations of the country of origin.

Hodder & Stoughton Ltd
Carmelite House
50 Victoria Embankment
London EC4Y 0DZ

www.hodder.co.uk

Contents

We wrote this book together as a joint work, but each chapter is written either from Sarah's perspective or from Cath's. For clarity, we have indicated the author of each chapter at the top of the page throughout the book.

Note on naming conventions		ix
Prologue: 'We don't know what's in it'	(Cath)	1
Chapter 1: We Made a Vaccine	(Sarah)	5
Chapter 2: Disease X	(Sarah)	29
Chapter 3: Designing the Vaccine	(Cath)	57
Chapter 4: Money, Money, Money	(Sarah)	77
Chapter 5: Making the Vaccine	(Cath)	97
Chapter 6: Scale-up	(Cath)	125
Chapter 7: With Great Care, and Due Haste	(Sarah)	149
Chapter 8: Trials	(Cath)	163
Chapter 9: The Prince and the Protestors: Vaccine Acceptance and Hesitancy	(Sarah)	189
Chapter 10: Vogue	(Cath)	219

Chapter 11: Waiting (Sarah) 235
Chapter 12: To Licensure and Beyond (Sarah) 259
Chapter 13: Disease Y: Next Time (Cath) 275

Acknowledgements 293
Appendix A: Different types of vaccine 299
Appendix B: The classic method and the rapid method 305
Appendix C: What is in the Oxford AstraZeneca vaccine? 311
Notes 315
Index 325

Determination and perseverance move the world; thinking that other people will do it for you is a sure way to fail.

Marva Collins

Better to light a candle than curse the darkness.

Anon

Note on naming conventions

The virus, the disease and the vaccine that are the subjects of this book all acquired and/or changed their names during the time the book covers. We have adopted the following approach.

For the virus: novel coronavirus then SARS-CoV-2. The virus was officially named in mid-February 2020.

For the disease: novel coronavirus then Covid-19 or Covid. The disease was officially named in mid-February 2020.

For the vaccine: ChAdOx1 nCoV-19 then AZD1222. Also: Oxford vaccine, Oxford AstraZeneca vaccine, AstraZeneca vaccine, Covid vaccine, Covid-19 vaccine.

The vaccine was officially renamed AZD1222 after we partnered with AstraZeneca.

The UK's Medicines and Healthcare products Regulatory Agency (MHRA) and the EU's European Medicines Agency (EMA) called it ChAdOx1-S [recombinant]; the Serum Institute of India who are manufacturing the same vaccine in large quantities call it Covishield; and in April 2021 it acquired another official name, Vaxzevria, but we do not use any of these names in the book.

PROLOGUE

'We don't know what's in it'

I spent the first weekend of August 2020 on a camping trip with my 9-year-old daughter Ellie and some friends. The last six months in Oxford had been crazy, but now here we were, on a remote campsite in the north-west of Wales. With our little tent pitched by a mountain stream, no phone signal and no electricity, it was a welcome chance to get away from it all and take a breath.

By the Sunday evening Ellie had had enough of my campfire cooking and we'd decided to treat ourselves to dinner from the pizza van. Wandering over to the van, I saw that my friend Gabi and her Jack Russell Cookie had befriended another woman-and-dog combination. Their conversation had started like so many in the last few months: 'How have you been coping?' 'Aren't these times strange, and difficult?' 'At least we can get out into the countryside.' 'Maybe everything will be back to normal by Christmas.'

I started complaining about the lack of phone signal (I'd enjoyed the peace at first, but now my mind was on the Zoom meeting I had the next morning) and the conversation moved on to our new friend's worries about the ongoing 5G installations across the country.

I offered that Public Health England had found the technology to be no risk to health, and I didn't see why it should be contentious to use a bit more of the radio spectrum to give us better connectivity. After all, we all used mobile phones, and we all knew 4G was better than 3G. She was unconvinced: 'I just worry,' she said. 'We just don't know.' But we *do* know, I thought.

As the dogs continued to sniff each other, Gabi said with a laugh, 'At least you're not saying 5G causes Covid. Or that Bill Gates is using the pandemic to fit us all with microchips.'

There was a pause. Gabi had touched a nerve.

'I'm not saying there is definitely a conspiracy,' the woman put forward, carefully. 'But I do worry that we don't know what they put in these vaccines: mercury and other toxic chemicals. I don't trust them. They don't tell us the truth.'

At this moment, our pizza – tomato and salami, no cheese, at my daughter's insistence – was ready. But it would have to wait. Of course I wasn't away from it all. Why had I imagined I would be? It was everywhere, and affecting everything.

'OK,' I said after taking a second to think, 'I have to introduce myself. My name is Cath Green and I might not look like it in my bare feet and this dress – I might not sound like it either, believe me I know – but I *am* "them". You couldn't have known this, but I'm the best person in the world to tell you what's in the vaccine. I work with the people who invented it. It's me and my team, in my lab, who physically made it. We ordered the ingredients, we made the first batch, we made more batches from that, like with a sourdough starter, we purified it down and we put it into the tiny little vials. And what's in those vials is what's being used now in the trials. You say you don't know what's in the vaccine, but I do. I know exactly what's in it, and you can ask me anything you want about it.'

We talked for fifteen minutes or so: about my team at Oxford University; how the vaccine was designed; and how we were trying to make enough of it for global supply, to make sure that, once it's been shown that it works and is safe, everyone who needs it can get it. I went through the list of ingredients (which is also set out at the back of the book). I told her I would absolutely be prepared to give it to my own family.

It was all very amicable, and I hope that I told her enough about me to let her see that I am not the thing she worries about: a global elite, out to win power and control. I don't have Bill Gates's phone number. I don't know how to put a nanobot tracker into a vaccine. I'm just Cath, a lighterman's daughter, doing my best with the knowledge I have and the people I work with, and missing hugging my parents like everyone else.

I don't know whether I changed our new friend's mind – and I'm sorry that I can't remember her name, or her dog's – but I hope our conversation let her see another side to the story: one that definitely has its moments of high-stakes human drama, but that's not sensational and isn't designed around newspaper headlines or tantalising clickbait. It's a story of decades of painstaking preparation, of collaborative teamwork across disciplines, sectors and nationalities that makes you proud to be human. It's a story of ordinary people coming together in extraordinary times to attempt an extraordinary thing.

I had many similar conversations that year with many different people, but it was by a pizza van in a field on the edge of Snowdonia that I decided I wanted to write this book. This was the moment I knew that we, the Vaxxers, needed to come out of our labs and explain ourselves. The story isn't over yet. The ending has not been written. But we have come a long way towards beating this virus and I would like people to know how we really got here and what happens next: how we get out of this mess, and how we prepare for the inevitable next one.

But, right then, I was on holiday – my daughter was getting bored and the sun was coming out. We said a friendly goodbye and I headed back to the tent, for a slice of cold pizza and a warm glass of wine.

CHAPTER 1

We Made a Vaccine

23 November 2020

This is the story of a race. Not, as it has so often been portrayed, a race against other scientists making other vaccines. With billions of people needing to be protected, we were always going to need all the vaccines we could get: ideally made using different technologies, so that if one failed another could step in; with different raw materials, to minimise the likelihood of global shortages; and in different countries, to protect against hoarding and vaccine nationalism. Instead, it was a race against the devastating virus that took millions of lives, ruined livelihoods, emptied schools, kept us apart from people we loved and closed down entire societies. It was a race that, even as the world locked down in the first months of 2020, we had in some ways already lost. It is a race we are still running, as mutant variants threaten to 'escape' the vaccines and treatments we have developed to bring the pandemic under control. And, if and when we do cross the finishing line, sadly there will not be much time for celebration. We will already be in training for the next one.

—

My colleague Cath and I decided to write this book in the summer of 2020. By this point, we had designed our vaccine – known as the Oxford AstraZeneca vaccine – made it, and started trialling it. We were fairly confident that it would work, but we did not yet have the data to prove it. The country was starting to open up again, and though there were warnings that there might be a 'second wave' people's lives seemed to be moving back towards something like normal. Our own lives were a little bit more normal too. We were occasionally having a day off at the weekend and I was forcing myself to get some exercise. We realised that we wanted to do something to engage with people's concerns about vaccines and show the exquisite care and attention that goes into making them safe. We wanted to dispel some myths about vaccines, science and scientists. And we wanted the chance to tell the real story, at least our part of the story, of how we made the Oxford vaccine.

Scientific discovery on this scale is very rarely a eureka moment for a lone genius. It definitely was not in this case and we hope we never sound as though we think we did what we did on our own. It was a collaborative effort by an international network of thousands of heroes – dedicated scientists in Oxford and across four continents, but also clinicians, regulators, manufacturers, and the brave volunteer citizens who offered up their arms for us, and week after week stuck cotton buds down their throats. And though there was plenty of drama there was not one big breakthrough moment, in a bath or under an apple tree or late at night in a silent empty lab, but rather lots and lots and lots of small moments. Detail after detail that we had to get completely right, item after item to be ticked off the list, problem after problem that had to be solved. Some of those were scientific problems: how do we make this genetically stable? What dose of the vaccine do we give for the best protection?

How many doses and how far apart? Others were economic, political, logistical: how can we check our volunteers' temperatures when thermometers are impossible to get hold of? How can we get vaccine made in Italy into the arms of volunteers in the UK when there are no flights? Where is the money for all of this coming from? Countless incremental small steps.

Some of the most important moments had actually happened before anyone had ever heard of Covid-19. Because whenever you are working at the cutting edge of science you are building on decades of meticulous and laborious work that has come before. The flipside of that of course is that, if we had been better prepared, we could have gone even faster. Those of us working in the field had expected something like this for years. At the beginning lots of people were asking 'why did we not see this coming?' The answer is that we did see it coming, and we had started preparing, but we had not been able to persuade anyone to spend the money that we needed to do what was required. So, what do we have to do to be ready for next time? What lessons must we learn to prevent us from having to go through this again?

We had always known that whilst vaccines had probably saved more lives than any other scientific discovery, they also, more than almost any other branch of medicine, made people anxious. Perhaps it has something to do with the needles. Perhaps it is that, whereas most medicine is given to someone who is already ill and makes them better, vaccines are given to people who are healthy, to prevent something that might never happen. Certainly, fears and doubts often rush in to fill gaps in people's knowledge. So, we wanted to try to fill in some of those gaps ourselves. What is a vaccine and why do we need them? How do they work? How has it been possible to make them so fast? What is in them? How do we know they are safe?

In 2020, a coronavirus causing a disease called Covid-19 swept around the world, wreaking havoc with healthcare systems, economies, and all of our lives. By the end of the year, it had killed more people in one year than any infectious disease had done for over a century. This book is the story of how two scientists were in the right place at the right time to fight back. We are not 'big pharma' or 'them'. We are two ordinary people who, with a team of other hard-working, dedicated people, did something extraordinary. We don't have cleaners, or drivers, or nannies, and like everyone else we had other things going on in our lives. There were days when we swore or cried with frustration and exhaustion. We lost sleep and gained weight. There were days when we met a prince, or a prime minister, and other days when it seemed that we had to both save the world *and* get the central heating fixed. Some days we drank champagne, others we struggled to find anything to eat for lunch. There were days when we seemed to be battling against our employer, or the media, or a swarm of wasps, as well as the virus. Most days, it felt like our big chance to make a positive impact on global health. Occasionally it felt like a heavy burden to bear. But we kept going, as did many others who worked alongside us, for long days, through weekends and bank holidays, until our vision of a vaccine for the world was finally realised.

—

For the rest of my life, people will ask me how I felt at the moment that I heard about the trial results for our vaccine. The answer is that I didn't have any strong emotions. Relief that the vaccine worked, certainly. Surprise that it was such a complicated result: three numbers instead of one – trust a bunch of academics to come up with that. Concern about what was going to happen next. And then I took myself off to bed, before anyone got the chance to talk to me: I wasn't allowed to discuss this with anyone.

It was late in the evening on Saturday 21 November. Professor Andy Pollard, my colleague running the clinical trials, had called me that morning to tell me the analysis of our trial data would be happening that weekend. Like the rest of the world, we had been waiting impatiently for this moment for weeks.

We had designed the vaccine in a few days in January (at that point, on a just-in-case basis); made the first batches in a record-breaking sixty-five days as it had become clearer and clearer that a vaccine would be needed; tested it on volunteers in four continents as the virus rampaged across the planet; and manufactured millions of doses. We already knew it was very safe. We had great confidence by this time that it would work. But we still did not have the data to prove it. The analysis we were waiting for would crunch tens of thousands of data points from the thousands of volunteers involved in the trials, to show whether, and how well, our vaccine had worked to protect them against Covid-19. Anything over 50% would be considered a success. When the analysis was done – probably sometime on the Sunday, Andy thought – that would trigger the process of informing key people and filling in the blanks in the press release that we had already prepared. Andy told me to 'get a good night's sleep, have a glass of wine', and wait to hear from him with the news.

I had first talked to Andy about him being chief investigator for the clinical trials back in February. My work is in early vaccine development whereas Andy had been responsible for several very large vaccine trials and had enormous experience in vaccine policy – how vaccines are actually used in the real world. At the time, Andy had no idea of the phenomenal amount of work he was letting himself in for.

With a busy week coming up I knew what I needed to do: laundry. I had also ordered a few new shirts for this occasion but trying them on now it was clear that one did not suit me at all and would need to be returned. Though this wasn't urgent, I felt that completing a task, however small, would be good for my mental health. The drop-off point was a newsagent's a mile away, and a walk on that crisp November day would also help.

As I walked, I saw quite a few Christmas decorations up already – on 21 November! After such a horrendous year it seemed people were getting their lights out early, in an attempt to cheer themselves up. It made me think of my children's first few years at school, when Christmas preparations would start from late October, with a Christmas play, meaning late nights after school, and then more talk of Christmas non-stop until finally term finished and they had several days at home being tired and not knowing what to do with themselves, and *it still wasn't Christmas*! This year, the early Christmas trees seemed to drum home the point that we were approaching the end of the year and *we still didn't know if our vaccine would work.*

I continued busying myself with small tasks. I couldn't even talk to my family about what was happening. But over the last year they had become used to me becoming non-communicative when things were difficult at work. They knew we had been under terrible pressure for the last two weeks, with blood samples from our 24,000 trial volunteers arriving in the lab and needing to be processed. One shipment of samples from Brazil had been due to arrive at 6.30 p.m. that Tuesday but had actually not arrived until 9.30 p.m., so people had been in the lab checking, thawing, dividing up, labelling, repacking and shipping until the early hours of Wednesday morning. Later that day I had made the unpopular decision to send home anyone not working on the blood samples: because of rising case numbers there were strict rules again about how many people were allowed to be in the labs. On the Friday we had started running out of freezer space, so I had spent the afternoon begging colleagues in other buildings for space in lockable freezers. On the Saturday morning, I had spent some quiet time in my office sticking labels onto tubes ready for the next shipment from South Africa. A lot of people were feeling stressed that week and so my family probably put my withdrawn behaviour down to all of that.

That evening I was trying to read but actually dozing off when my phone pinged me awake with a message from Andy asking me to do a video call. It was odd. I hadn't been expecting to hear from him until the next day, and I had assumed he would then just be telling me a number: the vaccine's efficacy is x%. Couldn't we do it over the phone, I messaged back? No: he needed to show me some slides.

I was still half asleep but also my heart was racing. Why was this happening a day early? Why were slides needed?

Once I had the laptop set up, Andy walked me through his slides. He quickly and calmly got to the point. Overall efficacy was 70%. The number wasn't as high as Pfizer's 90% or Moderna's 95%, announced earlier that month. But it was better than the 50% that was deemed the minimum for the vaccine to be useful, and much better than the 30% that some commentators had been warning we might need to be prepared for only weeks earlier. Our vaccine was effective.

But it was more complicated than that.

In the trials we had tested a number of different dose patterns. While the analysis showed an *overall* efficacy of 70%, intriguingly, in the group of volunteers who had received a half-dose followed by a standard dose, efficacy was 90%. In the group who had received two standard doses it was 62%.

Whenever scientists are presented with data that is unexpected, we wonder if it is a chance finding – a statistical blip. But looking at it in more detail, this didn't seem to be the case.* Colleagues at AstraZeneca would now spend Sunday repeating the analysis our Oxford University statisticians had already done so that we would have two independent analyses of the data. Because AstraZeneca is a publicly listed company, the results had to be announced by press release before the financial markets opened on Monday morning.

I spent most of Sunday in the office, in Zoom meetings with various groups of colleagues working on the analysis and the media plan, and occasionally popping into the lab to check that everything there was under control. But I couldn't chat for long because I couldn't let anyone in the lab know that I knew the efficacy result, let alone what it was. Everyone working on the vaccine had been warned that they would probably hear the results on the news. To give myself something to do I walked to the shops and picked up some fruit for the lab team. It had been pizza and cake for several days.

By 11 p.m. the wording for the press release – which I would need to approve – had still not been finalised. I gave up waiting and went to bed, slept fitfully, and woke soon after 3 a.m. to see that it had arrived in my inbox. I sent my approval and then tried to get back to sleep. The press release

* For one thing, the efficacy of standard dose/standard dose was 60% in the UK trial and 64% in the Brazil trial: effectively the same. (Half-dose/standard dose had not been tested in Brazil, only in the UK.) For another thing, while the main efficacy result was for mild symptomatic cases, we also had data on asymptomatic infection. For that, the half-dose/standard dose pattern was again more protective than the standard dose/standard dose pattern, meaning that in a second independent data set we were seeing the same result.

would go out at seven and I knew a day of media interviews awaited me.

Monday morning was frosty. At 6.30, knowing the puddles on the unlit cycle path would have frozen, I decided to abandon the bike, scrape the ice off the car, and drive in. In the otherwise empty atrium of my building two people were cleaning, wearing masks tucked under their noses, which is completely useless. I politely told them that the masks had to go over their nose as well.

Sitting in my office for what felt like the millionth time that year, I turned on my computer and clicked the link to the press team's spreadsheet to find out which journalist I had to speak to first. I couldn't get the spreadsheet open. After a lot of clicking I gave up, went to the loo with my hairbrush and make-up bag, and discovered what looked like a large insect bite on my left cheek, red and swollen. By the time Tess came into my office, around 7.45, I was starting to feel a bit tearful. Tess – Professor Teresa Lambe – is an immunologist and long-time colleague and friend of mine. She had designed the vaccine with me in the first days of January and been working as hard as me on it ever since. When she found me in my office that morning she thought I was welling up with emotion over the enormity of what we had achieved. She reassured me that that was understandable. On the contrary, I told her, it was sheer frustration. I had had very little sleep and now had to face a day of press interviews. It turned out that my first call was at 8.30, so there had been no need for me to leave home so early, scraping my knuckles on the frozen car windscreen. I hadn't had any breakfast either. I made myself a cup of coffee and as the caffeine started to kick in I took a few deep breaths and pulled myself together.

Then the day began in earnest. There was a video call with Prince William, who had given the team a huge boost when he had visited earlier in the year. Then it was an online press conference, where most journalists were focused on trying to understand the half-dose/standard dose complication (which we ourselves did not yet fully understand – we had known about it for less than forty-eight hours). Then multiple in-person interviews to journalists with film crews in the basement seminar room, followed by a series of telephone interviews. The second journalist to interview me asked how sick I was of the press and I said, 'Pretty sick.' He said he sympathised as he realised that people like me were having to talk to the press on top of their day jobs. I didn't think it was worth explaining that the Covid vaccine wasn't my day job – doing this was already on top of my day job, which was developing vaccines against five other diseases.

At some point in the afternoon someone popped their head around my office door to suggest buying champagne for the team members who were in the building. I was so tired I had really hoped to leave the celebrations for another day, but others were keen to have a socially distanced drink. The plan was for people to collect a glass from a table in the corridor, take it back to their desk, and join a video call. So at around five o'clock I poured myself a tiny bit of champagne. Sipping from my coffee cup (we were short of glasses) I started to relax. I briefly enjoyed that end-of-the-week winding down feeling, before realising it was still Monday. The week seemed to have gone on for several months already.

Some of the journalists had asked me what it had been like to tell my family, and I had to confess that I hadn't. I had just sent a WhatsApp message around 7 a.m. saying 'big news day'. When I got home, they hugged me and directed me out of the kitchen, where a celebratory meal was being prepared. We ate together, raising a glass of wine in a toast, and finally I was able to go and sleep.

—

That was Monday 23 November, seven months to the day since we had immunised our first volunteer in the first trial of our vaccine, and less than a year since anyone had ever heard of a novel coronavirus causing pneumonia in Wuhan, China.

There were some jubilant headlines, tempered by increasing concern about rising cases.[1] A lot was made of the fact that our vaccine, unlike others, did not need to be kept in ultra-cold freezers but could be transported and stored at normal fridge temperature. This would make it vastly easier to distribute and use, in the UK and around the world. A lot was made too of how much cheaper this vaccine would be, and the commitment to supply it at cost in perpetuity to low- and middle-income countries. There was excitement that we had 100 million doses on order in the UK, enough to vaccinate every adult, and billions of doses being manufactured around the world for 2021.

But it didn't take long before the negative press started.[2] Commentators and stock-market analysts in the United States attacked our results: we had not published enough information; we needed to be more transparent; we were cherry-picking our data; our half-dose trial results were based on a mistake and therefore suspect; we had not included enough older people in our trial. Anthony Fauci, since 1984 the director of the National Institute of Allergy and Infectious Diseases and one of the most highly respected scientists in the field, compared our efficacy results unfavourably with Pfizer's and Moderna's and asked why anyone would want to receive our vaccine.[3] The UK media started to pick up and repeat what was being said in the US.[4]

All of this felt very unfair, not to mention unhelpful: the important point was that we had a vaccine that was very safe and highly effective against an unbearably awful disease. But that point was getting lost. It had not been our choice to put the results in a press release. As academics, we would usually publish our full results in a detailed peer-reviewed paper, setting out clearly our methodology and full of graphs and Kaplan–Meier curves, for the whole scientific community to pore over. But we were working with a publicly listed company, AstraZeneca. And when a publicly listed company receives information that could affect its share price, legally it has to release that information straightaway, to prevent insider trading. So it had to be a press release first. The academic paper, which was submitted for peer review the day after the press release, would contain the details to answer all the criticisms being levelled at us, but until it could be peer-reviewed, edited and published, the information vacuum was being filled by speculation. We could only be grateful to those British scientists coming to our defence, and wait. As celebratory gifts arrived in my office – champagne, cards, flowers, chocolates, bottles of vodka, and a cheque for £10 from a grateful pensioner 'to buy some mince pies' – I wondered whether they had been sent before the negative press started, or in spite of it. In any case, we really appreciated the gestures.

A week later, cycling in to work, I was feeling somewhat sorry for myself, which is not a particularly constructive emotion, so during the privacy of my twenty-minute cycle ride, I asked myself why. I was getting endless requests for media interviews, which would only increase once our paper was published. I had to keep the plates spinning on all my other projects. I was feeling the pressure of having kept my knowledge of the efficacy results away from my family, which had made me feel cut off from them. And I missed my son. My daughters were at home, having reluctantly decided to study remotely this year, but I hadn't seen my son since September. Almost twenty-two years earlier, I had given birth to triplets, becoming a mother of three within the space of sixteen hours, and now having one of my children away from home in such difficult circumstances left me with an underlying sense of tension. I found myself checking the Covid case numbers for Bath, where he was living, every day.

For one interview, I had been sent a list of sixteen questions to prepare. The last one was 'What have you put on hold during 2020 that you are looking forward to doing when this is finished?' Even thinking about that as I read the email, my eyes filled with tears. I told them that if they asked me that question I would cry. The wall between my professional self and my private self was crumbling. So I prepared an answer that I could give without getting emotional: 'I want to take my family on a wonderful holiday.'

And of course, having worked so hard for such a long time, it was disheartening to be the subject of so much criticism. The best approach was to ignore it, but some of my colleagues seemed to be taking things personally. They wanted to engage point by point with the critics, and were keen to barge in and explain to me and Andy that by not doing so we were being 'defeatist'. I spent a lot of time explaining why their preferred approach was not going to be helpful.

Two long and difficult weeks after our initial press release, on 8 December 2020, the *Lancet* published our paper.* It was the first publication of full phase III efficacy data for any of

* The references for this and the other papers we published on our vaccine are in the notes section at the end of the book.

the Covid vaccines. But although our paper did answer the criticisms, it did not silence the critics. The US biotech press in particular was presenting vaccine development as a competition that required winners and losers, with Pfizer and Moderna cast as the winners and Oxford AstraZeneca as the losers. One article that stung referred to the 'middling performance' of our vaccine and 'fumbled clinical trials'.[5] In reality, the only competition was between a virus and human ingenuity. With my colleagues, I had been saying to the press for months that the world was going to need multiple vaccines; it wasn't about winners and losers, it was about saving lives.

—

I had anticipated that once our efficacy results were published things might quieten down. In fact, the next few weeks – as our vaccine was given a licence by the UK regulator, and then started to be rolled out in the UK and across the world – were if anything busier, and more of an emotional roller coaster, than ever. Whilst efficacy is of course an important number, the real *impact* of a vaccine is dependent not just on that one number, but also on how much of it you can make (supply); how many people you can get it to (delivery); and how many of those people then take it (acceptance). And apart from that, what matters in the long term is not 'efficacy', as measured in a clinical trial, but 'effectiveness', meaning what happens when the vaccine is used on a large scale in the real world. Unless vaccines are compared head-to-head in the same clinical trial, the efficacy figures cannot be directly compared because of sometimes major, sometimes subtle differences in the way the trials are set up and run.

On the day that our *Lancet* paper was published, the UK became the first country in the world to begin to roll out the Pfizer vaccine. But supply was limited and the vaccine needed very careful handling, and special arrangements for ultra-low temperature storage. Initial roll-out was slow, with 86,000 vaccines administered in the first week, and around 600,000 in the following weeks.[6] At the same time, case numbers in the UK soared dramatically, at least partly fuelled by a new, more transmissible variant of the virus – known then as the Kent variant, the English variant, or B.1.1.7, and now as the Alpha variant. Tighter and tighter government-mandated restrictions were being announced at a dizzying pace; NHS staff, dealing with rising case numbers and concerned about falling ill themselves, were asking to be given priority over people living in care homes for the limited supplies of vaccine available.[7] And then there was a last-minute admission that we would not, after all, be able to go ahead with plans to relax restrictions over Christmas.

But in the midst of this, good news did come. On 30 December, the UK regulator, the Medicines and Healthcare products Regulatory Agency (MHRA), announced emergency-use licensure of the Oxford AstraZeneca vaccine. It was a triumphant day for the team: the culmination of almost exactly a year's work. That morning I listened to the six familiar pips of the BBC time signal and then the headlines on the *Today* programme: 'Good morning. The headlines. The Oxford vaccine has just been approved for use. Tier 4 restrictions are to be imposed on millions more people across England to counter an upsurge of coronavirus infections. MPs are expected to approve the post-Brexit trade deal.' Lying there hearing our achievement summed up so briefly, I reflected that it was months since I had still been in bed at 7 a.m., and hoped it was a sign that normality could be returning after a turbulent and frantic year. Exultant headlines once more: this was a 'bright shaft of light', a 'game-changer', 'in the worst of times . . . the best of news'.[8] With so many more doses available, of a vaccine that was so much easier to handle, vaccinations could be speeded up dramatically.

The following Monday, 4 January 2021, after the world had marked a very subdued New Year, the roll-out of the Oxford AstraZeneca vaccine began. As a medical doctor working on the clinical trial, Andy was one of the first in line. Since I was not a healthcare worker or in any of the other priority groups, I would have to wait my turn. That evening, with case numbers and deaths continuing to soar, and the NHS on the verge of being overwhelmed, the prime minister, Boris Johnson, put the country back into lockdown. Schools and non-essential retail would be closed again, and mixing between households banned. The clinically extremely vulnerable were once more asked to shield themselves. Our vaccine was being rolled out into a grim situation and for a few days the media seemed determined to seek out negatives that only added to the sense of desperation.

For example, it had been decided to change the interval between doses from three or four weeks to twelve weeks. This was in order to give as many people as possible a first vaccination, given the extraordinarily high case rate and the dire situation in hospitals. It was a complicated decision to make: we had data showing higher immune responses with our vaccine with a twelve-week interval, but there was no comparable data for the Pfizer vaccine and a lot of people were very upset about it.* There was also a flurry of misinformed reporting about mixing

---

* The decision was made on the advice of the Joint Committee on Vaccination and Immunisation (JCVI), a group of experts whose role is to advise political decision-makers in the UK on immunisation strategy. In the UK, as elsewhere, there is a regulatory body that gives approval for use – the MHRA – and a policy body that assesses how a vaccine should be used in the light of prevailing circumstances – the JCVI. The MHRA had licensed the vaccine for use at two doses given at an interval

vaccines.* Next came questions about what was preventing 2 million people or more from being vaccinated every week, with the government trying to blame vaccine supply, which did not fit with the information I was receiving from AstraZeneca. There were 'postcode lottery' stories about some parts of the country having no vaccination centres. And there were worries about the new variants of the virus – then known as Kent, South Africa and Brazil and now as Alpha, Beta and Gamma – and whether the vaccines announced in such triumph over the last few weeks would prove effective against them.

After a few days, while news about the pandemic, the pressures on hospitals, and daily deaths remained absolutely grim, news about the roll-out became more positive. Buckingham Palace announced that the Queen and the Duke of Edinburgh had been vaccinated. More vaccination centres opened, including several cathedrals and, importantly, to counter false claims that the vaccine was not suitable for Muslims, a mosque in Birmingham. The Serum Institute in India started supplying vaccines through the region, including to Myanmar, Bangladesh and Nepal.

---

of four to twelve weeks. The JCVI advised that it should be given with a twelve-week interval. Later, in Europe, their regulatory body – the EMA – licensed our vaccine for use in adults over 18 but national policy bodies such as Germany's STIKO then made various different decisions about use of the vaccine in different age groups.

* Advice was issued saying that in some circumstances people could receive one of the licensed vaccines as their first vaccination and the other one as their second. This was completely misreported as a recommendation to do that, which it was not. It was simply saying that in the situation where someone had received a first vaccination and arrived at a clinic to receive their second, and either it was not known which vaccine they had received first, or doses of that vaccine were not available, they should receive the vaccine that was available rather than being sent away. I was receiving a lot of emails from people saying that I should protest, because this was untried and would be unsafe. In fact, I have worked on mixing different vaccine technologies in this way for more than twenty years and I know that not only were their fears unfounded but this is frequently more effective than two doses of the same vaccine. In February 2021 we announced a clinical trial to gather data on this approach.

Then the roller coaster took a series of very sharp and unexpected turns. Politics, business and science came together in an emotional, stomach-churning mix. As the date approached for the European regulator to make a decision on our vaccine, Pfizer, Moderna and AstraZeneca all announced temporary issues with supply of their vaccines to the continent.

The EU's vaccination programme was going very slowly. Now there was slightly hysterical talk from the EU of an export ban to prevent Pfizer vaccine intended for the UK from leaving the bloc; there were inspections of an AstraZeneca factory in Belgium to ascertain whether the vaccine being made there and intended for the EU was being 'siphoned off' to the UK; and there were demands that vaccine made in the UK and intended for the UK should be commandeered by the EU. A German newspaper published a completely inaccurate story saying that the AstraZeneca vaccine would not be licensed in the EU for over 65s because it was 'only 8% effective' for this group.*[9] President Macron of France weighed in by saying, again completely inaccurately, that the AstraZeneca vaccine 'seems quasi-ineffective on people older than 65'.[10] The whole thing was like the joke about the two crotchety old women in a restaurant. First crotchety old woman: 'The food here is terrible.' Second crotchety old woman: 'Yes, and such small portions.'† Except here the complaints were about our vaccine. 'This vaccine is terrible.' 'Yes, and they aren't giving us enough of it.' The

* After a small media storm it transpired that the German journalist's source had perhaps confused the number of people in the trial aged 55–69 (8%) with the efficacy of the vaccine for people aged over 65. Although it was a bit more complicated than that. It was also true that there was not a great deal of data to prove efficacy in older people at this stage. But there was a great deal of data to strongly suggest it, and this was a pandemic situation.

† I thought this was an old Jewish joke, which it may be, but it also makes an appearance in the opening monologue of Woody Allen's *Annie Hall*.

European regulator did, that same day, and as expected, approve the vaccine for all adults.* That was a huge news day for vaccines, with two more, Novavax and the single-dose Janssen/Johnson & Johnson, announcing excellent efficacy results similar to ours.[11]

Over this period, several major vaccine developers left the field, or announced they were going back to the drawing board, for various reasons. After the stunning success of the first three vaccines to report results – Pfizer, Moderna and our Oxford AstraZeneca vaccine – it was a stark reminder that this success had not been inevitable. It was disappointing to hear that these other programmes from leading vaccine developers had run into problems. A range of different vaccines, using different technologies and produced by different companies in different countries, were our best chance of getting a vaccine to everyone who needed one in 2021. I did occasionally allow myself to feel a bit sore, though, that we were continuing to get bad press for our successful vaccine, while others were receiving sympathy for their unsuccessful attempts.

After a slow start, the UK's vaccine programme now quickly ramped up. On 3 January, the day before the start of the AstraZeneca roll-out, and nearly a month after roll-out had started, 1.4 million doses had been administered. By the end of January, less than a month later, this number was 9.79 million, with a record 609,010 given on 30 January. By 15 February the government had met its target of offering a first dose to the nearly 15 million people in the first four priority groups – the most vulnerable, and the health and social care workers taking care of them.

* Some European countries, including Germany, made the policy decision not to use it for older adults. In Germany, they perhaps felt that they had plenty of Pfizer and Moderna vaccines and so would only approve use in the age groups for which there was clear evidence of efficacy, which we did not yet have, though we were confident we soon would.

Meanwhile the vaccine was being licensed and distributed around the world: by the end of January it had been approved for use in numerous countries including Brazil, Chile, India and South Africa.* On 15 February it was approved for use by the World Health Organization (WHO) which was hugely important for global roll-out and access to the vaccine for low-income countries. Within a couple of weeks, the first shipments arrived in thirty more countries including Ghana, Senegal, Rwanda, the Democratic Republic of Congo, Cambodia and Moldova. By 23 April, one year to the day since the first vaccination had gone into the first volunteer's arm, the vaccine had reached 172 of the world's 195 countries, from Afghanistan to Yemen.

As more data became available, we were able to show that our vaccine was actually more effective than our initial analysis on 23 November had suggested. And that a twelve-week interval between doses provided better protection than a shorter interval, vindicating the decision that had, of necessity, been taken without the benefit of a full data set. It also became clear that our vaccine would not only protect those vaccinated from becoming ill, but would also significantly reduce transmission of the virus to others. Our vaccine was shown to provide protection against the Alpha variant, and would most likely prevent at least severe disease caused by the Beta and Gamma variants. All of this was extremely good news.

And then on Monday 22 February, the first 'real-world effectiveness data' was published for the two vaccines being rolled out in the UK, Pfizer and AstraZeneca. Three months on from our phase III trial efficacy results, this is a moment I

* The full list at this point was Argentina, Bahrain, Bangladesh, Brazil, Chile, Dominican Republic, Ecuador, El Salvador, Hungary, India, Mexico, Morocco, Myanmar, Nepal, Pakistan, the Philippines, Saudi Arabia, South Africa and Thailand.

doubt I will ever be asked about, but it is one I will always remember.

A study of the entire population of Scotland, including 1.1 million people who had received a vaccine, found that the Pfizer vaccine was 85% effective in preventing hospitalisation three weeks after a first dose, and the Oxford AstraZeneca vaccine was 94% effective. It also reported that in people over 80, effectiveness of both vaccines after a first dose was 81%, which was the first time I had seen data from that age group reported.[12]

I read these numbers with wide eyes and a sharp intake of breath. It's common for effectiveness data – meaning data about how a vaccine works when it is used in the whole population in the real world – to be lower than the efficacy data reported in clinical trials. By way of illustration, in our trial, no one who had received the vaccine had been hospitalised, giving an efficacy against hospitalisation of 100%. In real life, with the vaccine being given to mainly very elderly people, some of whom were in care homes or were already suffering from failing health, the number would never be 100%. But to see 94% was incredible. This was the crucial first data on the impact of our vaccine on deaths, hospitalisations and illness out in the real world, and it was spectacular. And in the first direct comparison of two licensed vaccines in large numbers of people in the real world, our vaccine was performing as well as Pfizer's, despite the differences in the efficacy data reported from clinical trials.

There was still a lot for us to do – we were still very busy working on vaccines against new variants, and it was concerning that our vaccine was still unpopular and mistrusted in continental Europe. But it felt like our roller coaster might at last be slowing down and allowing us to step off, tentative, blinking and a bit shaken, back to normality.

The last weekend of February 2021 was bright and sunny, and I finally stopped using the excuse that the fields were too muddy to run in, and took myself out for a jog. In my garden, sparrows were collecting nesting materials and shrubs that had been bare twigs were now covered in green buds. There was a sense across the whole country of the mood lifting. The vaccine roll-out continued to be hugely successful and there was increasing confidence that it was reducing infections, easing the pressure on the NHS, and saving lives. The prime minister had announced our route out of lockdown, beginning with the return of children to school. The possibility of spring, of emergence from the darkness of the pandemic winter, started to feel real.

We had relaxed too soon. The roller coaster had at least one more loop for us. Again the venue was Europe and again it involved an unedifying mix of politics, business, science and emotion.

However, on Monday 22 March, with the use of AstraZeneca still suspended in several European countries due to concerns about some rare health issues that may or may not have been caused by the vaccine, and the looming threat of a vaccine war that would be disastrous for global supply, the long-awaited interim results from the US trial came out. They found that the vaccine was well tolerated with no safety concerns;* that it was 79% effective in preventing symptomatic Covid; and that it was 100% effective against severe disease. In older participants, efficacy was just as good, at 80%.† AstraZeneca had run the trial in the

* 'Well tolerated' is the term used by clinicians, researchers and regulators to indicate that there have been no significant side effects that would prevent us using a vaccine – in everyday language it is safe.

† These numbers were updated soon afterwards to take into account further cases. The overall efficacy number went down from 79% to 76% and efficacy in adults over 65 went up from 80% to 85%.

US because US regulators are reluctant, even in a pandemic, to rely on data obtained outside the US. But in the meantime the US was doing well rolling out vaccines from Pfizer, Moderna and Johnson & Johnson. Despite its determination to pursue an 'America First' approach under President Trump, the US had now inadvertently provided a gift to the rest of the world: data that would, I hoped, restore confidence that had been eroded by careless words and political gestures.

There will inevitably be more loops on the roller coaster, more bumps in the road. Science doesn't exist in a vacuum. When we started to work on our vaccine, across the whole team we were drawing on years of experience in vaccine design, production, clinical trials and regulatory affairs. Between us we had in-depth experience of so many aspects of what would need to be done, and the will to do it quickly, always planning multiple steps ahead, with early vaccine roll-out and public health benefits in mind. We wanted to save lives, not to make money. And we wanted to draw on everything we had learned from the previous attempts to develop vaccines at speed so that we didn't waste time waiting for someone else to play their part.

What none of us foresaw was how the vaccine would become a political football. Our year of constant, painstaking attention to detail resulting in a vaccine with the potential to save millions of lives around the world could be dismissed by a politician with a grudge. Carefully worded statements to the media, explaining the science behind the vaccine, would disappear in a Twitter storm of bias and misinformation, with incorrect statements repeatedly cited as fact. Were we simply too idealistic? Too naive? Certainly next time – and there will be a next time – we should add some political scientists to the team.

Stepping back to reflect, I would remind myself that for years people had been telling me the technology behind my vaccine was too slow, or too expensive, or just not good enough. Less than a year earlier I had been struggling to raise a few hundred thousand pounds to begin work on this vaccine. Just a few months earlier, sensible scientists had still been saying we might never develop an effective vaccine against Covid-19. Just a week before Pfizer had reported its 90% efficacy results, we were having our expectations managed that 50% would be a good result and 30% might be acceptable.

In fact, in less than a year, we had produced a vaccine that was very safe, that was highly effective including in older people, that prevented deaths or hospitalisation after one dose, that reduced transmission, that could be transported and stored at fridge rather than freezer temperature, and that would be available around the world in huge quantities at low cost. We had made a vaccine for the world.

CHAPTER 2

Disease X

1–10 January 2020
Confirmed cases: 4–59[1]
Confirmed deaths: 0

On New Year's Day 2020, I was sitting at my desk at home, going through my work emails and browsing a few news websites. I checked in with ProMedMail, a site that reports on disease outbreaks around the world, and something caught my attention. There were reports of 'pneumonia of unknown cause' in Wuhan, China. Four cases with high fever and pneumonia, not responding to antibiotics. First patient worked at a seafood market. Interesting.

Further down the page it was suggested that this could mark the return of SARS, but that 'citizens need not panic' and more information would follow. Pneumonia of unknown cause could be lots of things and I made a mental note to check back later on these four cases. For now, it was still the holidays, so I shut down my laptop and went to join the rest of the family in the kitchen, where we had a new Christmas jigsaw on the go, and a stack of half-completed crosswords.

When I looked again the next day there was a new report.

Twenty-seven people, most of them stallholders from the seafood market, had been hospitalised, and seven were in a serious condition. The market had been closed.

Transmission from seafood to humans seemed highly unlikely to me: fish and shellfish are not normally considered to be sources of viruses that can infect humans. However, reading on, I saw that there was now confirmation from local media that other animals, including pheasants, snakes and rabbits, had also been on sale in the market.

By 3 January, in between a cold sunny walk through the Oxfordshire countryside, a pub lunch, and making dinner to suit each family member's different taste, I was checking in regularly to look for updates. A post that day reported forty-four cases, with eleven people critically ill, and 121 close contacts of those infected being monitored. The disease was 'SARS-like', the post said, but it was still unclear what was causing it: whilst influenza, avian flu and adenovirus had been ruled out, there were many other respiratory viruses that could be responsible.

I checked the website again that evening, looking for new details – new clues – and I ran through the possibilities in my mind. Maybe whatever it was would simply disappear: we might learn what it was, but never need to produce a vaccine against it. On the other hand, this really could be a new outbreak of SARS. Or something 'SARS-like'. Or something even worse.

—

As a professor of vaccinology at the University of Oxford, I had been working for years on developing vaccines, most recently against so-called emerging pathogens. Pathogens are micro-organisms that can make us ill. Emerging pathogens are those pathogens that don't normally infect humans – although we may come into contact with the wild or domesticated animals that they do infect; and that have the potential to cause severe outbreaks when humans become infected; and for which few or no vaccines or treatments exist. So my job is to develop vaccines against some of the emerging pathogens we already know about, and prepare to make a vaccine very quickly if a new one is identified.

If this new illness in China was indeed SARS, or SARS-like, we were in trouble. In 2002, a virus previously unknown to science and now known as SARS or SARS-CoV (severe acute respiratory syndrome coronavirus) began to cause pneumonia in a province of China. Scientists came to the conclusion that it had most likely first been transmitted from bats to humans via small mammals like raccoon dogs, ferret badgers, civets or domestic cats. The disease spread, eventually infecting 8,000 people from twenty-nine countries and causing 774 deaths. There was no vaccine or specific treatment and around 10% of those infected did not survive. Vaccine development got underway, but it didn't get far before the disease had been contained and eradicated by the centuries-old but very effective methods of contact tracing and quarantine: find the cases, identify their contacts, and keep the contacts in quarantine until it is known that they are not infected. The outbreak had started in November 2002 and was all but over by June of 2003. There was no immediate need for a vaccine, although we couldn't be certain that there would never be another outbreak of SARS. The virus was presumably still circulating in bats, and maybe other species.

Another very nasty coronavirus then appeared in 2012. As it started infecting people living in the Middle East, it became known as Middle East respiratory syndrome, or MERS-CoV. Coronaviruses are often found in bats, and many simply stay within bat populations without ever causing any problems for humans. However, sometimes – as had happened with SARS, and as probably also happened in this case with MERS – the bats can pass on their infection to other mammals to which humans are more commonly exposed. It eventually became clear that MERS-CoV was endemic (meaning it was common) within camel populations in the Middle East, and that people coming into contact with infected camels were themselves becoming infected.

For many people living in societies in which camels have played an important role for centuries this was hard to accept. Instead many other animals were considered as the source of the infection. Could it be cats, or dogs? What seems most likely to have happened is that around fifty years earlier camels from Africa already infected with MERS-CoV from bats were imported into the Middle East. The disease causes barely any symptoms in camels so it was not obvious that they were bringing a viral infection with them. Gradually, though, the infection then spread amongst the camels in the region.

MERS in camels is a lot like the common cold in humans. Young camels don't become severely ill with it, but they do spread it. Camels develop antibodies as they recover, but the antibodies do not provide lifelong protection, so they can be re-infected later, again experiencing only a very mild or even asymptomatic illness.

If MERS had remained in the camel population it would not have caused a problem and probably would never have been discovered. However, as more and more camels became infected, it spilled over into humans, meaning that people started to become infected as well. In Africa, where the original infected camels came from, camels tend to be kept in herds like cattle or sheep, but in the Middle East camels can be treated more like family pets, with much more contact with humans, and so more opportunity for the virus to be passed on.

As we know, coronaviruses are passed from human to human when tiny droplets breathed or coughed or sneezed from the nose or mouth of an infected person enter the body of another person, through their nose, mouth or eyes. The MERS virus in the breath of an infected camel is breathed in by anyone coming close to that camel. But that alone doesn't necessarily mean the virus infects that person. There's another crucial step before a virus can cause an infection.

In order to infect a cell, a virus needs to latch onto a receptor on the surface of that cell. The cells that make up our bodies are surrounded by a smooth membrane, which is studded with receptors that sense the environment around the cell and allow it to interact with signals coming from other parts of the body. For example, adrenaline, produced in response to stress, fear or exercise, has to bind to a particular receptor on the surface of a cell to exert its effect. Different viruses latch onto different receptors, and the fit has to be good for the virus to be able to infect the cell, a bit like a phone charger needing to fit the socket on a phone. For example the MERS-CoV virus latches onto the dipeptidyl peptidase 4 protein, known as DPP4. It isn't important for our purposes here to understand much about DPP4 (it plays a part in immune regulation); it is only important to know that it turns out that DPP4 in camels and humans is almost identical. This is the key to how the virus could infect humans so easily. Had this receptor been different in humans, the virus that easily infects camel cells and is transmitted between camels would have found it difficult to infect a human cell. The camel virus could have been breathed in with no effect on a person at all. There are millions of viruses that for this reason are no threat to us whatsoever. However, in this case, MERS-CoV happens to bind onto DPP4 on both camel and human cells very efficiently. So the virus passes easily from camels to humans.

By January 2020, around eight years after the first MERS infection was found, there had been over 2,400 confirmed cases in twenty-seven countries, and 858 deaths. Most of the cases had been in Saudi Arabia, where there had been outbreaks in older people and in hospitals. Once the virus gets inside a hospital, where there are many susceptible people in close proximity, it can spread quickly. This is exactly what happened in South Korea in 2015. A South Korean businessman had travelled in the Middle East and became unwell after returning home. In the nine days between initially seeking medical help and diagnosis, he visited a number of hospitals, resulting in an outbreak that infected 186 people, of whom thirty-eight died. More than 6,000 people were quarantined to bring the outbreak under control, thousands of schools were closed and there were significant effects on the economy as retail sales declined and interest rates were cut.

The South Korean outbreak was closely studied as it offered a rare opportunity. Since it had been initiated by one known individual it was possible to track the course of the outbreak very precisely and gain an understanding of how the disease had spread. Healthcare workers were tested, and virologists also tested samples of the air and surfaces in hospitals to find out where the live virus could be found. What they discovered was that even young healthcare workers had become infected and, despite having very mild, almost asymptomatic infections, they had been able to spread the disease to others. This made MERS quite different from SARS – with SARS infected people always had recognisable, often severe symptoms, making the job of contact tracing and quarantine somewhat easier.

While SARS has been eradicated, MERS continues to cause

infections in the Middle East to this day. To try and keep infections down, the Riyadh camel market has now been moved well away from the city, as I discovered myself when I visited it with Naif Alharbi, a Saudi former student of mine. But this has not been enough to stamp out MERS. Not all of those infected turn out to have had direct contact with camels; meanwhile some young and healthy people who work with camels evidently have antibodies against MERS but are not aware of ever having been ill with it. It seems likely that young and healthy people working with camels are becoming infected and suffering a mild, cold-like illness, during which time they are unwittingly transmitting the virus to others who are more vulnerable.

So, you may be wondering, what does all of this have to do with the reports of an illness in China in January 2020? In order to understand how much of a problem was likely to stem from whatever was causing illness in China, we needed to understand certain characteristics of the disease. The two key questions were: is it caused by a virus that is being transmitted between people; and if so, are people able to pass the virus on before they show signs of the disease themselves? While the suggestion that this new infection could be SARS, or SARS-like, was definitely a cause for concern, the same ProMedMail post on 3 January 2020 also said that no human-to-human transmission had been observed and no medical staff had been infected.

One possible scenario that fitted the facts was that this was a previously unknown virus that infected an animal, like rabbits. Maybe, I puzzled to myself, a stallholder had brought to the market a few cages of apparently well but actually infected rabbits. Perhaps, as with MERS, the rabbit virus could infect people who spent time in close proximity to an infected rabbit. With MERS, though, only older people had become seriously ill, and most outbreaks had happened in hospitals. With this new virus, this 'pneumonia of unknown cause', it appeared that, in contrast, younger, healthy people were falling ill and there had been no outbreaks in hospitals. So perhaps, unlike MERS or SARS, this virus was not able to spread from human to human. Perhaps all the people falling ill had been at the market and were being infected directly from the rabbits. If that was the case, then the outbreak would be contained very quickly by closing the market, removing all the livestock, and deep cleaning the site before reopening. There would be no time to develop a vaccine, nor any need for one.

That scenario fitted the facts as I understood them. But of course, we did not yet have all the facts. Passing the kitchen table that evening, I tried putting another couple of pieces into the Christmas jigsaw. They didn't fit. I would come back and look at them later.

—

In order to understand what we now all know intimately as Covid-19 and to explain how my colleagues and I designed a vaccine to protect against it, we first need to travel 12,500 kilometres west from Wuhan in China to Guinea in West Africa, and six years back in time. In 2014 there was an outbreak of an extremely deadly and terrifying disease, Ebola. Ebola had first been identified in 1976 and there had been numerous small outbreaks since, but this one was much larger and was running out of control. During the 2014 outbreak – which eventually killed at least 11,000 people, caused some hospitals to become overwhelmed and close down, and had grave economic consequences for the region – I played a peripheral role in the testing of an Ebola vaccine. But from my seat on the sidelines I was able to follow developments closely and see what went well, and what did not.

The inadequacies of the world's response to this disaster, and the determination of the scientific and public health communities to learn from those mistakes, are key to understanding what would happen next. These experiences formed the foundations of what we've done in Oxford – because of what happened with Ebola, learning what worked well and what did not, we were able to design, make and test a vaccine with unprecedented speed and with high levels of confidence in the outcome.

Ebola is a filovirus. Whereas coronaviruses like SARS, MERS and SARS-CoV-2 cause respiratory tract infections – from colds, coughs and sore throats to pneumonias and breathing difficulties – filoviruses cause haemorrhagic fevers. Ebola symptoms are initially much like those of other viral infections: fever, headache, fatigue and muscle ache. After a few days, though, this can progress to vomiting and diarrhoea, problems with the liver and kidneys, and internal and external bleeding. It's a truly devasting disease: it is fatal in around 40–50% of those infected.

Before 2014, outbreaks had been small, in isolated areas with low population densities, so it had been possible to contain them through painstaking contact tracing and quarantine. What was different in 2014 was that the outbreak spread to densely populated cities, and across borders, from Guinea in West Africa to the neighbouring Liberia and Sierra Leone. As the disease spread exponentially in the summer and early autumn that year, the healthcare system was overwhelmed and people were dying, untreated, in the streets of Monrovia, Liberia's capital city. By the time the outbreak had finally been contained in mid-2016 there had been more than 28,000 cases and over 11,000 deaths.

At the time there was no licensed vaccine against Ebola, and no treatments. However, this disease was so frightening that the US government had begun funding work on a vaccine to prevent the possibility of Ebola being released deliberately in a bid to cause death and panic. Over many years, various Ebola vaccines had been produced in laboratories and tested on monkeys, so that by 2014, two different vaccines had been shown to protect rhesus macaque monkeys against Ebola virus after a single vaccination.

One of these promising vaccines was known as ChAd3 EBOZ, which is an example of a *replication-deficient recombinant simian adenoviral-vectored vaccine*. This is not as complicated as it sounds if we just take one part at a time, and it is important for understanding what we did in 2020 with Covid-19, so let's break it down.

First, *adenoviral-vectored* – this means the vaccine is delivered into the body using an adenovirus. Adenoviruses are a group of viruses that cause colds in humans: snuffly nose, or sometimes gastro symptoms, mild and short-lived. There are lots of slightly different versions, known as serotypes, each identified by a number. Some of these human adenoviruses have been used to make vaccines, including Ad5 and Ad26.

The adenoviruses that give us a cold infect a cell in our nose or throat, and turn the cell into a factory to make lots more adenoviruses. This is known as replication. Eventually the cell breaks open, releasing the adenoviruses that it has made, which then infect more cells in the same person, or are coughed or sneezed out and may infect someone else. The virus spreads around the body, or to another person, or both. If a particular gene is removed from an adenovirus, this creates an adapted version of the virus that still infects human cells as normal, but is then unable to spread any further. The adapted adenovirus can't make copies of itself and spread around the body or to anyone else: it is rendered *replication-deficient.**

Where the gene has been removed from the adenovirus, another gene is then inserted in its place. This creates a *recombinant* virus, which simply means it combines genes from two different sources. The added-in gene is 'expressed' in the cells infected by the adenovirus – meaning the gene's instructions are carried out in the infected cells. The added-in gene instructs the cell to make a particular protein. An immune response is then triggered against that protein.

The *replication-deficient recombinant adenovirus* is live, which is good for inducing strong immune responses, but it is also very safe to use: being replication-deficient means that it cannot spread to any more cells and so will not cause an ongoing infection.

This is the technology that was used in the Ebola vaccine, and the same technology can be used to express a gene for an Ebola, or malaria, or influenza protein, and in each case it will induce an immune response against that pathogen. One way of thinking of it is to see the viral vector as a vehicle, able to deliver whatever cargo is loaded up into it. Once the cargo (the gene for the Ebola protein, for example) is inside the human cell, the virus uses the human cell to make lots of copies of the Ebola protein, rather than lots of copies of itself. The immune system, which continually scans the body looking for things it doesn't recognise, then springs into action.

---

* Why do I talk about 'replication' rather than 'growth'? Viruses do not grow and divide, the way bacteria or yeast do. The first part of that sentence comes with a vivid memory for me. In my first year at university, in a microbiology lecture, the lecturer wrote on the blackboard (it was a long time ago) VIRUSES DO NOT GROW AND DIVIDE, and told us, with considerable emphasis, that this should be written on our brains in letters of fire. What they do instead, in order to make more of themselves, is to take over a living cell and turn it into a factory for assembling and making more viruses. The correct word to describe this process is replication, not growth.

An Ebola vaccine made using human adenovirus 5 had been produced and tested, and it had worked really well in animals. The problem was that the vaccine was made from an adenovirus that quite often infects humans in its unmodified state, and so lots of people already have antibodies against it. In those people, the vaccine wouldn't induce such strong immune responses against the added-in Ebola or malaria or influenza protein. The idea of using an adenovirus to produce a vaccine was good, but it would be better to use an adenovirus that people didn't already have antibodies to.

The problem could be neatly solved by using an adenovirus that circulates amongst chimpanzees rather than humans. There are many different adenoviruses found in chimpanzees that are perfectly capable of infecting humans and giving them a cold. But because humans don't often come into contact with chimps, we haven't been infected by these chimpanzee, or *simian*, adenoviruses and so we don't have antibodies against them.

The Ebola vaccine ChAd3 EBOZ was made from a chimpanzee adenovirus known as ChAd3. The E1 gene was removed to render it replication-deficient, and the added gene was one that encoded the protein found on the surface of Ebola virus, the glycoprotein. When used to vaccinate macaques, the macaques produced immune responses against the Ebola glycoprotein. With these immune responses they were protected against Ebola virus.

So this was one of the vaccines that had been shown, by the time of the 2014 outbreak, to protect macaques from Ebola. The other vaccine was another recombinant viral-vectored vaccine, so it used a very similar technology. The main differences were that it used a different viral vector – a virus called VSV rather than the adenovirus – and it was replication-competent, meaning it could spread throughout the body after vaccination.

By 2014, both vaccines had been shown to protect macaques

against Ebola virus but neither had undergone human trials. As the outbreak in West Africa rumbled on, it became obvious that this was not just a West African problem. Having worked on influenza vaccines, I was well aware that a new strain of that virus could travel around the world in a matter of days. People in West Africa had already travelled in cars, lorries and buses, unaware that they were infected with Ebola, and taken the virus with them. With the virus now circulating in cities with international airports, it was only a matter of time before someone boarded a plane. Viruses do not respect borders, religious beliefs or political leanings. The wealthy might be able to shield themselves more effectively but even they are not safe from a catastrophic collapse of society. The whole world was susceptible. This was a global problem that required a global response.

An effective vaccine could control the outbreak and play a major part in bringing it to an end. Along with other colleagues at the Jenner Institute, my institutional home within the University of Oxford, I became involved in efforts to start a clinical trial of the ChAd3 EBOZ vaccine as soon as possible – in other words, trials to test the vaccine on humans.* In order to get to the stage where we could test it on people, though, we needed all of the necessary funding and approvals – something we will come back to in other chapters.

The timeline over the next few months was a large part of what went right, and then what went wrong, in our response to the Ebola outbreak. At first, we managed to move unusually fast. Then things got bogged down. We submitted our grant application to fund the clinical trial on 14 August 2014. By 17 September, after a massive amount of careful work by clinicians,

* We had not been involved in the development of ChAd3 EBOZ up until then, but the only previous clinical trial of a ChAd3-vectored vaccine had been done in Oxford, which was why it was helpful for us to get involved.

scientists, and regulatory and ethical bodies, approvals were all in place and we were able to vaccinate the first of sixty volunteers in Oxford: Ruth Atkins, a 48-year-old NHS manager and former nurse. Moving at such speed involved a lot of work by many people both at the Jenner Institute where I was based and at the Clinical Biomanufacturing Facility where Cath is now based, in close consultation with the regulatory body. In that sense it was a small taste of what was to come in 2020. We were delighted by our rapid progress, which was necessary given the public health emergency that was unfolding. By October 2014, with results exactly as we had hoped and expected based on our existing knowledge of how adenoviral-vectored vaccines behave, there was sufficient data on safety and immune response to allow a second study in Mali to go ahead.

Mali was not one of the main countries affected by the outbreak but it shares a border with Guinea and there was an expectation that Ebola might spread over the border. So it made sense for the trial to be conducted on healthcare workers who would have had to respond to any Ebola cases. In this trial the vaccine was shown again to be well tolerated with no safety concerns and to induce strong immune responses. However, no one yet knew whether the immune response would be strong enough to provide effective protection against Ebola. For that, efficacy studies in the affected countries would be needed, to see if the vaccine actually prevented people from being infected and becoming ill. Now, frustratingly, things slowed right down.

Trials to test vaccine efficacy are called phase III trials. They follow on from phase I or first-in-human trials, which test a vaccine's safety and ability to produce an immune response in young healthy adults; and phase II trials, which test for safety and immune response in a wider range of age groups. Typically all of these trials are blinded randomised placebo-controlled trials

– this means that half the people being vaccinated receive the vaccine, and half receive a placebo injection, which will not provide any protection. The placebo might be simply saline solution, which will not result in any immune response, or it might be another vaccine such as rabies vaccine, which won't protect against Ebola but will provide some other benefit to the volunteers who don't get the Ebola vaccine. Neither the volunteers themselves, nor the people assessing them later on to find out if they have become infected, know which vaccine they have had. When enough people on the study have become infected the statisticians 'unblind' the study and count up how many infections there have been in the group that got the Ebola vaccine and how many infections in the group that got the saline or rabies vaccine.

If there are the same number of infections in both groups the vaccine is not effective at all. If none of the infections are in the Ebola vaccine group, the vaccine is extremely effective. If there are some infections in the Ebola vaccine group, but fewer than in the other group, then the vaccine efficacy can be calculated. Usually if the vaccine efficacy is 70% or better, it will be worth using the vaccine to control the outbreak.

In this case, a placebo-controlled study was deemed unethical by those planning the trials. Ebola has a high fatality rate at around 40–50%, and there is no cure. How, it was argued, could anyone conduct a trial in which people had a 50:50 chance of getting a vaccine which might protect them, and a 50:50 chance of getting one that definitely would not?

In fact, the fatality rate had been dropping, as more healthcare centres were able to provide better supportive care for those infected and increase their chances of recovery. Anyone taking part in the trial would be closely monitored and, if infected, would receive care as early as possible, thus improving their chances further. And, the longer discussions about ethical trial design went on, the longer no one was receiving a vaccine that gave them any chance of protection at all. People continued to be infected, and people continued to die.

However, there were other options for a phase III trial. Another type of trial is a step-wedge design. This is a pragmatic approach, that takes account of the fact that if the outbreak is widespread and you want to introduce a vaccine to control it, that cannot happen everywhere at the same time. So instead vaccinations start at one edge of the outbreak area and gradually move across it over a period of weeks. If the vaccine is effective, infections in the area where people have been vaccinated will decline first, and the fall in infections will follow the progress of vaccinations across the outbreak area. Everyone receives the trial vaccine as the vaccination teams reach their area, although those living furthest away from the start point have to wait. Although this sounds like a possible solution, with this set-up it is actually quite difficult to work out how effective a vaccine is. Getting a clear picture of a vaccine's efficacy is especially important if there are two options, as there were for Ebola, and it would be useful to know whether one is more effective than the other.

So instead, another type of trial design was eventually decided upon: a ring vaccination study with delayed vaccination in half of the rings. In order to use this type of trial design it is necessary to first identify someone who has been infected with Ebola. Then all of that person's contacts are identified, and the contacts of their contacts, and the limits of the geographical area where they can be found are defined. That group of people forms a 'ring' around the initial case. Many rings are identified in this way, and each ring is randomly assigned to receive either immediate vaccination or delayed vaccination. That means that everyone in all of the rings receives the Ebola vaccine, but some rings have to wait three weeks. The number of new infections in the 'delayed' rings is compared with the number in the 'immediate' rings to assess whether the vaccine is effective.

Whilst all of the discussions about how to conduct the phase III trials had been going on, so had the Ebola crisis. It was frustrating to see the process slow down dramatically when a vaccine was so desperately needed. However, intensive efforts to bring the virus under control using contact tracing, quarantine and hygiene measures were paying off, and by April 2015 when the ring vaccination study started – a full year after the outbreak became widespread – the case numbers were low and falling. Originally the aim and expectation had been that the two vaccines would be tested for efficacy at the same time in different locations. By this point, though, there was only one location with high enough infection levels for a phase III trial, so only one vaccine could be tested at a time. VSV was to go first, with a switch over to ChAd3 later on.

The results of the VSV trials showed it to be very highly effective. This was great news. The vaccine was then used to bring the remainder of the outbreak under control: rings continued to be identified, and people in all rings were vaccinated straightaway. In October 2015, the WHO recorded the first week without any new cases and in June 2016, after some final flare-ups, the epidemic was declared over.

However, with the VSV vaccine bringing the outbreak under control, the ChAd3 vaccine was never tested for efficacy. Comparisons showed that the immune response was very similar for both vaccines, so it is likely that ChAd3 would also have been highly effective in protecting against Ebola. It also has a number of potential advantages over the VSV vaccine: it is easier to manufacture in large quantities, it doesn't require very low temperature storage, and, because it is replication-deficient, it is safer for people with compromised immune systems.

For those of us looking on, having been involved in the very rapid initiation and completion of the first ChAd3 clinical trials in Oxford, this seemed to be a missed opportunity. Why did it take so long to test for efficacy? It was four months from the outcome of the phase II trials until the start of the phase III study. I made this point at a conference where I had been invited to speak about our vaccine trial, and received a rather angry response from a WHO representative who insisted that everything had been done as fast as possible. But the fault does not lie with individuals not doing their job properly in the thick of things. The problem was a lack of preparation.

The fact was that the delays meant not only that it took longer to contain the deadly Ebola virus, but also that only one vaccine ended up being tested for efficacy – a vaccine that required very low temperature storage, making it difficult and expensive to use in hot countries.

In order to know if any vaccine works, unfortunately there does have to be a 'control group' in which some people who have not received the vaccine become infected, in this case with a potentially fatal disease. But once a vaccine has been shown to be effective, it can then be used to prevent many more potentially fatal infections. Between December 2014 and April 2015, while the discussions about the design of the efficacy study were going on, around 5,000 people became infected and more than 2,000 of those people died. In the ring vaccination study that was eventually carried out, just twenty-three people became infected whilst waiting for their vaccination. In a randomised placebo-controlled trial, as in a ring vaccination trial, only a very small number of infections would have been required in order to calculate vaccine efficacy, and if the vaccine had been shown to be effective, those originally given the placebo could then have been offered the vaccine. If we had started randomised placebo-controlled trials for both vaccines in December 2014, at sites where the small number of cases that would need to occur in the control group could have been treated quickly, effectiveness would have been established sooner. Both the VSV and the ChAd3-vectored vaccine could have been tested, either at different sites or one after the other at the same site, and by early 2015 we would likely have ended up with two effective vaccines to deploy against this and future outbreaks, rather than just one.* Many fewer lives would have been lost in 2015.

In truth, despite having known about Ebola virus since 1976, and despite having invested in vaccine development over many years, the world was not ready in 2014 to test a vaccine against Ebola. The response was simply not adequate.

—

The world's inadequate response to Ebola was a wake-up call, and a turning point. Relief that the outbreak had been contained was tempered by anxiety that there were still plenty more viruses out there that could wreak similar havoc. True, outbreaks of these very frightening diseases had always been small and contained in the past, but that had been true of Ebola too until 2014.

Various international organisations, with the WHO foremost, began to draw up lists of other dangerous pathogens for which

* In 2021, there was another outbreak in Guinea, infecting at least eighteen people and killing nine by the middle of March. Close examination of the virus revealed it to be almost identical to the strain that caused the 2014 outbreak. It seemed likely that someone who had been infected during the earlier outbreak had not completely got rid of the virus from their body, and now new infections were being caused. Thankfully this time, vaccinations could begin quickly.

we really should be developing vaccines, diagnostics and therapeutics.* In 2016 the WHO published a list of priority diseases, including SARS, MERS and Ebola, as well as Lassa fever, Marburg virus, Crimean-Congo haemorrhagic fever, Rift Valley fever and Nipah virus. The idea was that companies and academic researchers could put forward their technologies for review and some would be fast-tracked so the world would be better prepared for future outbreaks.

For me this was great news – I had been working on a technology that had great potential, but lacked the funding for further testing. In Oxford, we had conducted safety trials for ChAd3 EBOZ, the Ebola vaccine that never got to phase III trials. This was a replication-deficient simian adenoviral-vectored vaccine. Since 2012 I had also been working on another replication-deficient simian adenoviral vector, developed in my lab in Oxford and known as ChAdOx1, which I had used to make vaccines against influenza and MERS. So far both had been shown to be well tolerated with no safety concerns in phase I clinical trials. However, without more funding we would not be able to make further progress.

ChAdOx1 is what is called a 'platform technology' meaning that it can be used to make many different vaccines. The great advantage of this is that you don't need to repeat every one of the many aspects of vaccine development for each new vaccine that is made. Knowledge of how to manufacture the vaccine, how to store it, and what dose to give, can be built up and applied to every vaccine that is made using the same platform. This reduces development time and, crucially for the development of vaccines against diseases like those on the priority list,

* Vaccines are ways of protecting people from becoming infected, diagnostics are ways of testing whether someone has become infected, and therapeutics are ways of treating the infection to cure it or ease its symptoms.

where there is so little funding available, it also reduces the development cost.

In February 2016, in a featureless conference room in a hotel near Geneva airport, I presented this particular platform technology to the WHO panel. After this first round of assessment I was encouraged to apply for a second round. I was also advised to form a partnership with a large company that was interested in similar technology. As a university, Oxford can supply the expertise that supports the early development and clinical testing of vaccines, but eventually, if its vaccine is found to be safe and effective and is going to be licensed and used, it will need to bring a commercial company on board. The company would then take the lead in obtaining the necessary regulatory approvals, carrying out the manufacturing at scale, and organising the logistics of supply. I opted to partner with a company called Janssen, which is part of Johnson & Johnson, who were also using a replication-deficient adenovirus to make vaccines.

My main collaborator at Janssen was Olga Popova, a very experienced project manager and a very impressive and efficient person. Russian by birth, Olga spoke perfect English, worked for a Belgian company with sites in the Netherlands, and lived in Rome with her French husband and young daughter. She was always smartly dressed and could tell you the best hotel to stay in for WHO meetings in Geneva (not the one near the airport) and the best thing on the restaurant menu. We went into the second round of presentations in July 2016 as a team and later that year our technology was approved.

While this was a success (being approved was certainly validation), the WHO didn't have any money to hand out. So we would still need to find some funding if we were to make any more progress with it.

Fortunately, the WHO was not the only organisation that had leaped into action after the 2014 Ebola outbreak. A new organisation was set up in 2017 called the Coalition for Epidemic Preparedness and Innovation (CEPI, pronounced 'Seppy'). CEPI would fund the development of the vaccines the world needed to protect us all against emerging pathogens, and in particular the priority diseases. It also wanted to reduce the time taken to develop new vaccines from a stop-start ten years to a much slicker twelve months.

The first target for CEPI was vaccines against MERS, Lassa fever and the particularly nasty Nipah ('the one you don't want to get' according to scientists who work on it). Working with teams both at Oxford University and at Janssen, I submitted a proposal to develop ChAdOx1-based vaccines against all three diseases. However, this time I was to miss working with Olga again – my efficient, stylish colleague was diagnosed with cancer and sadly passed away in July 2017.

The full proposal required a lot of work, and it did not help that the deadline kept shifting. We had a family holiday in Italy booked for the first two weeks in July, after the three children's A-levels. We would be spending a few days in Milan, then going to Lake Como, and on to Venice, before flying home via Milan again: plenty of opportunity for the 'boat trips' the family teased me I hankered after on every family holiday. Fortunately, the deadline for the application was at the end of June, so I was looking forward to going on holiday with a clear conscience, having submitted it. Then CEPI shifted the deadline to a week later, halfway through the holiday. Then it moved again. I eventually completed the application sitting on the steps of the campanile in St Mark's Square in Venice, typing on a BlackBerry, while my partner Rob and the children were climbing the tower to check out the view from the top.

Despite the last-minute Venetian editing, we were invited to move forward to negotiate a contract for CEPI funding. This was unknown territory for all concerned. It was the first time Oxford University and Janssen had applied jointly for funding, and it was unusual for CEPI to have two applicants to negotiate with. On top of that, CEPI was still a new organisation, and its funding came from a variety of sources, each with different conditions attached, all of which then had to be passed on to the projects being funded. The negotiations went on and on, with many meetings and teleconferences, until finally the contract signing took place at the CEPI headquarters in London in September 2018. It was eighteen months since the original deadline for proposals, back in March 2017, twenty-one months since the launch of CEPI, more than two years since the WHO call to speed up vaccine development. And we had achieved nothing beyond securing some funding. You may be starting to get an idea of why vaccines usually take so long to develop.

Fortunately, in the meantime the UK government had decided to channel some of its overseas aid budget into vaccines against emerging pathogens, and I had secured some funding for early development of a MERS vaccine, outside of any CEPI involvement. This turned out to be extremely fortuitous.

Also in the meantime, in February 2018 a new entry had appeared on the WHO's top-ten list of priority diseases: 'Disease X'. Disease X was a placeholder, representing a future, hypothetical disease. No one knew what the disease would be, or when it would emerge, but experts agreed that the emergence of something, sometime soon, was inevitable. Looking back now, it's interesting to see that then, the planning was for an epidemic, meaning an outbreak in one geographical area – but not a global pandemic. Nonetheless, the very act of putting Disease X on the list was an important recognition of the need to prepare for pathogens we don't yet know, as well as the ones we do.

So how could we prepare for a disease that we don't yet know about? How do you design a vaccine against an unknown pathogen? Up until the invention of platform technologies such as ChAdOx1 this would have been an unsolvable problem. But platform technologies were designed to be adaptable and could therefore be used to make vaccines against many different pathogens, potentially including one that hadn't been identified yet. The other advantage of a platform technology was that so much of the work – determining how to manufacture it, deciding what dosage to use – could be done before we knew what disease we wanted to use it against. But even with the time saved by using a platform technology, the development process remained painfully slow: typically in my lab there was a delay of several years between the decision to start work on a vaccine against a particular virus and the start of clinical trials. This was largely because each stage of development needed funding, which we could not even apply for until we had successfully completed the previous stage of development. A funding application could easily take a year to go through, and some funds were only open to applications for a small window once a year.

Despite these limitations, we had already started to think about what we could do at our end, in the lab, to speed things up at the very beginning of the process. So when CEPI put out a call for proposals for platform technologies for rapid vaccine development against Disease X, we put in an application. We had done a lot of the planning, and some of the preliminary work, on how to cut down development time, but because the next step would involve a lot of expensive work in our manufacturing facility – for example, doing all the testing of the materials to be used in any vaccine ahead of time, so that when a new pathogen came along we could simply 'plug it in' to the adenovirus and start manufacturing – we needed the kind of funding CEPI could provide to take things any further.

We were hopeful. There were other platform technologies that, on the face of it, were better candidates than ours, because they could go faster at the initial vaccine construction stage. But they each had their own drawbacks. Nucleic acid vaccines, for example, consisting only of DNA or mRNA* (a relatively new kind of vaccine) should be simple and quick to produce once the genome of a new disease is identified. However, mRNA is so unstable it needs to be formulated with lipids, making the manufacturing process more complicated. And even with the addition of lipids the resulting vaccine needs to be stored at very low temperatures. So whilst the development of mRNA vaccines against Disease X might be fast, manufacture, transportation and storage present challenges. An older technology, DNA vaccines, had been used in clinical trials (I had designed such a vaccine, and been vaccinated with it in a trial). They are well tolerated, but as things stand they don't induce very strong responses in humans: so, fast, but not very immunogenic.†

Adenoviral-vectored vaccines do induce strong immune responses even with a single dose; there were already well-developed manufacturing processes for producing these vaccines at large scale; and they do not need to be stored at very low temperatures. So we felt that if we could get CEPI funding to work on speeding up that early development part of our process, we would be in a position to create a very strong technology for making safe, effective vaccines against new diseases at pace.

---

* DNA is the chemical material that makes up our genes. DNA has to be copied into RNA before its instructions to the body can be carried out. This type of RNA is known as mRNA, with the 'm' standing for 'messenger'.

† In 2020, mRNA vaccines against Covid-19, from Pfizer and Moderna, were indeed developed fast, and did also come with the drawback of having to be kept at extremely low temperatures. No DNA vaccine against Covid-19 has yet got as far as a phase III trial.

However, in the end, our application was not successful. The reviewers apparently were not convinced that a viral-vectored vaccine could be produced quickly enough in the event of a Disease X outbreak.

It was disappointing, but not unexpected: for an academic in my field, or any field, a 30% success rate for grant applications is pretty good, and we become used to rejection. We mentally filed the ideas to return to later and continued with our work on MERS, Lassa and Nipah vaccines, plus new work on Crimean-Congo haemorrhagic fever. I was also working on influenza vaccines, and collaborating with colleagues developing vaccines against other diseases on the WHO's priority list: Rift Valley fever virus, Chikungunya, Zika, Ebola, and Marburg virus. For all of these projects we were using ChAdOx1, gradually building our understanding of this platform. We weren't able to do the work we wanted to on our rapid response to Disease X though, which was the cause of much regret once Disease X appeared, not so very long afterwards.

—

By those early days of January 2020, with my grown-up children still at home for the holidays, the Christmas jigsaw still out on the table, and those first reports of SARS-like pneumonia appearing, a lot of the pieces were already in place. A lot had been achieved before Disease X came onto the scene and devastated and derailed so many lives. We were reasonably well placed to respond to a possible new virus. We had carried out all of the necessary steps for designing, making and testing a new vaccine before; we had planned how to get off to a quick start, even if we hadn't been able to work on that as much as we would have liked; and, once we got as far as planning clinical trials, we had a lot of information about things such as how long we could store the vaccine and what dose to use.

Over the next couple of days it became clear I was not the only person following the reports out of Wuhan with interest. I had previously collaborated with two scientists at the US National Institutes of Health Rocky Mountain Labs, Vincent Munster and Neeltje van Doremalen. By 5 January Vincent and Neeltje were sending emails to make sure I was aware of the outbreak – 'now 59 infected, and not SARS or MERS' – and wondering when the genome of the virus causing the infection would be released.

Back at work in my office on Monday 6 January, I talked through it all with my colleague Tess Lambe. We discussed the continued lack of evidence of human-to-human transmission, which didn't mean that it wasn't happening, but simply that there was no clear evidence that it was. Very late at night on 8 January came official confirmation on ProMedMail that we were looking at a novel (meaning a previously unknown) coronavirus.

A number of epidemiologists immediately posted on ProMedMail to say 'I told you so'. And they were right, they had told us. As experts in this area they had been warning for some time that there was likely to be an outbreak of a novel coronavirus, most likely starting in China. On Friday the 10th, China reported the first fatality, a 61-year-old man. By the end of that Friday, Tess and I had decided that as soon as we could get hold of the genome sequence of the novel coronavirus we would start to make a vaccine. We would follow the same design approach we had used for our MERS vaccine, and go as fast as we could.

At this point it all still felt quite theoretical. But this was, at least, a chance to demonstrate what we could do with our

ChAdOx1 vaccine technology; and if a vaccine *was* going to be needed, moving quickly, at every stage, and planning multiple steps ahead, was the required approach. We had been thinking about Disease X. We had been planning for Disease X. Could this be it? Or would the outbreak fizzle out? A year from now, I thought at the time, as we waited impatiently for the coronavirus genome sequence to be released, perhaps we will have demonstrated we can go fast, but we will have made a vaccine that no one wants or needs.

CHAPTER 3

Designing the Vaccine

31 December 2019–31 January 2020
Confirmed cases: 0–9,927
Confirmed deaths: 0–213

I threw a New Year's Eve party at my house to mark the end of 2019. We dragged the sofa into the garden so that people could sit by the little firepit, danced in the kitchen with cocktails and fizzy wine, ate chocolate cake and 'small food' (our family name for canapés), hugged and laughed. At midnight, we gave the children 'kids' champagne' (Appletiser) and let them unleash a party-popper bombardment.

It was a bittersweet evening for me. My husband had left me a few months earlier and I was still angry and hurt and a bit fearful of what 2020 would bring, but a night of fun with my friends made me feel loved and alive, although very hungover the next day. Some of my friends turning the sound up louder and louder that night were fellow scientists from the university. Others were pub-going parents from my daughter's school.

No one mentioned a new virus in Wuhan, China. On New Year's Day I cleared up the empty bottles and ate pizza on the sofa with my daughter. It was just a calm family day.

On Monday 6 January I went back to work. I juggle two roles at the university: I work half-time as head of the Clinical Biomanufacturing Facility (CBF) and half-time running a research team at the Wellcome Centre for Human Genetics. The CBF, a vaccine-manufacturing facility serviced by twenty-five highly qualified, very dedicated scientists and engineers, is an unusual set-up for a university: there are only a few universities across the globe that have their own GMP (good manufacturing practice) manufacturing facility, which means we can manufacture medicines that are going to go into people.*

We have three main roles at the CBF. We make the 'starting materials' from which we or others can make clinical-grade vaccines (in other words, vaccines of a high enough quality to put into a human). We make clinical-grade vaccines themselves. And we also test, label, certify and release those vaccines for clinical trials. Our goal is to produce really high-quality innovative medicines to combat diseases that globally cause significant harm but from which no one is likely to make any profits. Pharma companies might throw money at heart disease, but nobody's throwing money at Ebola or Crimean-Congo haemorrhagic fever.

One of the reasons that I love my work is that vaccines are such cost-effective and harm-reducing public health interventions. In other words, if you are going to put money into trying to improve the length and quality of someone's life, your best bet is to give them a vaccine against a nasty disease. Even before Covid-19, vaccines were saving an estimated 2–3 million lives a year and enabling vastly more people than that to live healthier,

* GMP refers to the worldwide set of minimum standards that anyone manufacturing medicines for humans must meet in their production processes.

longer more productive lives.[1] The CBF team has made vaccines to protect people against a wide range of diseases over the last fifteen years. These include diseases caused by bacteria (including meningitis, tuberculosis and plague), parasites (malaria) and viruses (including Zika, rabies and MERS). For some of these diseases, there are no known treatments. For others, the treatments are often expensive, or difficult to administer in time in low-income countries.

Viruses are a fascinating phenomenon. They are not alive: they cannot reproduce themselves unaided, and when not inside a host cell they are wholly inert. But once they infect a host cell, they can almost completely take it over, converting the cell's efforts into simply making more virus instead of doing the things that it would normally do. This process usually kills the host cell. So viruses have to be transmissible; that is, they are able to infect different cells in turn, spreading from one cell to another. When you or I get exposed to a virus for the first time, we might be lucky. The virus might not manage to latch itself onto any of our cells and so we do not become infected. Or, we might be unlucky, in that the virus does latch itself onto and infect one or more of our cells. The virus then takes over those cells, and the cells stop doing what they need to do and start making copies of the virus, which harms our cells in such a way that symptoms of that viral infection occur, making us feel ill.

However, at the same time as the virus is infecting us, it is also being detected by our immune system, which has evolved to be able to detect previously unencountered intruders; determine that they are unwanted; and then inactivate and destroy them, in the process laying down a memory of the intruder so as to respond to it better the next time. And our immune systems (made up of so-called B cells and T cells) are generally very good at this. B cells produce antibodies, which bind to the outside of the virus and stop it from infecting our cells; and T cells are a second line of defence that can recognise cells that have become infected and destroy them.

The reason that viruses can nevertheless make us sick is that they are very fast. It is possible for a viral infection to take hold before the immune system has had time to mobilise an effective response. This is where vaccines come in. During a viral infection, your immune system tries to tackle the virus. If it succeeds, then you get better, and you are left with an immune system that now remembers what that virus looks like, has trained itself to fight it, and will spring into action much quicker if it sees it again. This is why people don't usually catch chicken pox twice. But you do need to get sick with chicken pox the first time, and that's not fun. Vaccines aim to give your immune system this memory of what a dangerous virus looks like, so that you can tackle it effectively if you come across it, but *without you having to get sick with the virus in the first place.* They achieve this by presenting a harmless mimic of the virus to your immune system.

There are various ways that vaccines can do this. The vaccinations against smallpox developed in the late eighteenth century by Edward Jenner used a related but less harmful virus, cowpox. Many traditional vaccines present the body with a weakened or inactivated version of the virus. Modern platform or 'plug and play' vaccine technologies use the latest understanding of how biology works to show the immune system only the part of the virus that it needs to recognise in order to produce an immune response. Usually this will be a protein on the surface of the virus, completely harmless on its own. At the CBF, the approach we have the most experience with is the adenovirus-vectored vaccine. We have been using this method to make vaccines for research colleagues like Sarah since 2007, and using the ChAdOx1 platform since 2012.*

I pretty much always feel like I have too much to do, and January is a particularly busy month. In early January 2020, the CBF team was hard at work on the production of a new vaccine against Ebola for our colleague Tess Lambe. We also had new team members arriving who needed to learn the ropes. They had no idea what was about to come at them – none of us did. But their fresh eyes, combined with the knowledge and experience of my existing team members, would prove crucial to the effort we were about to embark on. In my other role at the Wellcome Centre, I co-direct a PhD programme and in January I always have hundreds of applications to review and shortlist and lots of admissions interviewing to do. I also help out with admissions for the Nuffield Department of Medicine, and organise a big annual conference for the UK Genome Stability Network. In short, my diary was fully booked up with just the everyday stuff of my normal professional life for the first two weeks of the year.

I was aware, while I was juggling all of this, of the novel coronavirus in China that was occasionally featuring in the news. As the month progressed more and more people were raising the topic with me and asking whether I was working on it: my friends had all heard of the concept of Disease X, probably when I was boring them senseless in the pub after a long day at work. But I told anyone who asked that I wasn't working on it and I didn't think this was Disease X. I admit, that was how I saw it at first, and I wasn't alone. At this point, Wuhan had not locked down and the WHO had not declared

* There's a bit more about different approaches to producing vaccines in Appendix A at the back of the book.

an international public health emergency. Outbreaks like this happen all the time, I reassured my friends, and they usually die out locally. I was busy, I had other things on my mind, and I assumed this would be the same.

Luckily, others were not so preoccupied. Scientists in Wuhan, Shanghai and Beijing had started using advanced sequencing machines (essentially, machines that can 'read' an organism's genetic code) on swabs taken from patients with Wuhan's new, strange pneumonia. They were working to unravel the virus's 28,000-letter genome: the vital code that would tell the world what this virus was, how it was likely to behave, and how it could be stopped.

Sarah and her team were busy too, designing the vaccine. There was no reason for me to know about this – at this point, it was mostly a just-in-case exercise. It would be another few days of blissful ignorance before I realised that perhaps this *was* Disease X and perhaps I *was* going to be working on it.

—

Even before the genetic sequence of the novel coronavirus was published, Sarah and Tess had decided exactly what the vaccine would look like, and had started to discuss the method they would use to make it.

In the past, the design phase for a new vaccine would always have involved doing a lot of small-scale work in Sarah's research lab first. If, after testing a selection of research-grade vaccines in the lab and immunising some animals, she decided we had something promising enough to manufacture for clinical trials, she would then raise the funds to begin all over again making the clinical-grade starting material at the CBF.

This is because in the research labs, they don't use the very pure, very high-quality, very highly controlled ingredients we need to use when we manufacture for clinical trials. For example, in one of the early stages of the process of making a vaccine, we need to grow bacteria in a nutritious broth, or growth medium. The liquid used in research labs contains small amounts of animal-derived ingredients. Any animal-derived material used for a vaccine destined to go into someone's arm has to be very strictly controlled (in terms of its origins and how it has been processed), so the CBF uses an all-vegetarian alternative called Vegebroth, which not only is more expensive but is also a less effective medium for growing bacteria, so it slows down the work.

Making the vaccine in the research lab and testing it out before we start making it in the CBF is a bit like a dressmaker making a toile or muslin of a couture garment. Before cutting into the expensive material she will make several versions in a cheap fabric to check the fit. Once the fit is perfect, she will use the toile as a guide to make the catwalk version. Doing things this way can save money, because if the research vaccine doesn't work, we have not gone to the expense of making a clinical batch. It also means multiple vaccine options can be assessed in the lab, and the best one selected for further development.

This method had served us well for years, but it also of course builds in considerable delay. The whole process – applying for funding, making the vaccine in the lab, testing it, publishing the results, applying for more funding, then starting from scratch making the vaccine in the CBF – usually took at least three years. And that's just to get you 100 ml or so of starting material.

This time, Tess and Sarah knew that if they wanted to go quickly, they only had one shot. And they did want to go quickly. They wanted to show what their technology could do and how fast it could work. And they knew there was a chance it might actually be needed. So, they couldn't afford to waste even a few weeks testing different designs in their lab like they usually would before deciding which one to take forward to trials. That made the decision straightforward. (However, at this point I still didn't know anything about it because they weren't yet thinking about putting it into humans, which is where I come in.)

Most viruses have proteins studding their surface that become the target for vaccine development: it's the outside of a pathogen that the immune system needs to recognise. Flu viruses have two major proteins, so there is a decision to make about which one to target, but coronaviruses only have one kind of protein sticking out of them: the now famous spike protein. So, even before they had seen its genome, once they knew that this new pathogen was a coronavirus Sarah and Tess knew two crucial things: that the virus would have a spike protein, and that an effective vaccine would need to induce an immune response against that spike protein.*

Sarah had an obvious template to follow, because of the work she had already done on MERS, also a coronavirus. The MERS vaccine that had been made was a replication-deficient recombinant simian adenoviral vector vaccine, using the ChAdOx1 platform to deliver the gene for the MERS spike protein. We knew from years of research that vaccines made using the ChAdOx1 platform could generate a strong immune response with one dose and (because they were non-replicating) were safe to give to children, the elderly and anyone with a

* Almost all the 150 or so of the vaccines in development by the end of 2020 focused on inducing an immune response against the spike protein.

pre-existing condition such as diabetes. The MERS vaccine, specifically, had already been through two clinical trials and had worked well.

So Sarah also knew a third crucial thing. The design for the new vaccine would be exactly the same as the design for the MERS vaccine: essentially, the gene coding for the spike protein, plugged into ChAdOx1. We would take the pattern from MERS and go straight to couture.

The trickier question was, which method should we use to put the component parts together and start to generate the vaccine (which scissors and thread)? We had been making ChAdOx1-vectored vaccines in the lab and in the CBF for many years, but we were frustrated with how slow the very earliest parts of the process were, and had been wanting for some time to work on new, faster methods.

There were two circumstances in which we knew these faster methods would be important. Firstly, moving fast would be crucial if we wanted to make personalised vaccines against cancer. Immunotherapy to treat cancer is a growing field. Tumours often contain mutations in their DNA that make them different from the healthy cells in the body in a way that can be recognised by the immune system. In rare cases, people have mounted such a strong immune response against their tumours that they spontaneously recover. The idea of personalised cancer vaccines is to take a sample of the tumour, sequence its DNA, and then make a vaccine specifically for that patient that will immunise them against their tumour. In order to be of use to a cancer patient, the vaccine will need to be produced very quickly before the patient's condition deteriorates, although, since it is personal to each patient, not in large amounts.

Of course the other time we want to be able to go very fast is when we need to make a vaccine against a new virus at risk of causing an epidemic like SARS-CoV-2. Over the last couple of years, Sarah had invented a new, faster method for making the starter material for a vaccine. This new method – we called it the 'rapid method' as opposed to the 'classic method' – was all about trying to do as much work as possible in advance of knowing the genetic sequence of a tumour or pathogen so that, once we knew the relevant gene, we could get from that point to a working vaccine in someone's arm as quickly as possible.*

Think about a baker who sells personalised cakes iced with a message like 'Happy 50th birthday Joe', or 'Congratulations on your engagement Ali and Max'. She might wait until she gets an order, and only then start the process of mixing the ingredients, baking the cake, letting it cool, icing it all over, waiting for the icing to set, and then finally adding the customised message. If she gets the order the day before the cake is needed, that works well. But if she wants to be able to offer a quicker service, she could bake a stock of cakes and put on the base layer of icing every morning. She is taking a financial risk: if no orders come in, the pre-baked cakes will go stale and need to be thrown away. But it may be worth the risk. When a customer comes into the shop, all she has to do is pick up her piping bag and add the custom message while he waits. The cake is then ready to take straight to the party. Only in the case of a vaccine, the party is a pandemic.

In the classic, slower method, we make the cake (i.e. the starting material for the vaccine) from scratch when we get an order in (i.e. discover a new pathogen and establish its genetic code). In the rapid method, we make the cakes in advance, and when a new order comes in, all we have to do is add the last

* Appendix B at the back of the book describes the classic method and the rapid method in detail and explains how they differ.

bit of decoration (i.e. the appropriate string of DNA).* Doing the work in advance is an upfront investment of time and effort, but it is worth it in situations where time is of the essence.

Using the rapid method to make a vaccine against a novel coronavirus was obviously a great idea. But it was brand new. At the beginning of 2020, we hadn't ever actually used it. It was like a recipe that hadn't ever been tested: and now we had an order in from a customer who might turn out to be very important.

—

It was late on Friday 10 January. Tess and Sarah were still trying to decide whether they could use the rapid method or would be better off just using the classic method, when Chinese scientists made the genetic sequence of the novel coronavirus publicly available online.

A genetic sequence is a long list of the four letters of the genetic code – A, C, G and T – each of which stands for one of the four chemical compounds that make up DNA. And the sequence for this new virus was in Tess's inbox when she got up the next morning. She immediately started work (in her pyjamas: it was a Saturday).†

It was relatively easy for Tess to identify which sequence of letters amongst the 28,000 contained the genetic code for the spike protein, as the sequence was very similar to those for

* For neither method do we start making the vaccine with the whole pathogen itself. We never need to handle SARS-CoV-2 in the making of our vaccine. We just need to know the genetic sequence of the virus we want to immunise against, and then we order a piece of synthetic DNA that codes for the appropriate part of that virus – in the case of SARS-CoV-2, the spike protein.

† Tess says she regrets telling a journalist she was in her pyjamas and would like me to stress that she does usually get dressed to work.

the spike protein in other coronaviruses like SARS and MERS. However, over the weekend several different versions of the genome were posted. Tess had to compare them and make a decision about which one to use. That done, she sent the information over to another colleague, Sarah Sebastian, to do the final work on designing the DNA sequence.*

Sarah Sebastian works at Vaccitech, a spin-out company founded by Sarah Gilbert and others in 2016. It uses the same viral-vectored technology that we at the university use to make vaccines against emerging pathogens, but focuses instead on making vaccines to treat cancer and other diseases. Sarah S, originally from Germany and educated both there and in the United States, is a talented and very hard-working scientist. She had previously worked with Sarah G at the university on malaria vaccine development and improving viral-vectored vaccines: important work, but never something that funders get very excited about. Like many senior scientists, Sarah S had been finding the lack of job security working in an unloved field challenging. It's one thing to accept a short-term contract as a newly graduated researcher but she was experienced, and good at her job, and very reasonably wanted security for herself and her family. When she plucked up the courage to tell Sarah G that she had been offered a medical writing job, Sarah G, wanting to avoid a loss to the world of vaccinology, suggested that instead she should approach Vaccitech, who were looking to add to their team.

When we design a gene to go into a vaccine we don't

* In case you are confused because you know that coronaviruses are made up of RNA, not DNA, a short digression here. Most organisms' genetic code is made up of DNA but in fact coronaviruses contain only RNA as their genetic material. The four chemical compounds in RNA are the same as those in DNA, except that the T is replaced by a U. Adenoviruses use DNA, so Sarah needed to design a DNA sequence that corresponded to the RNA sequence coding for the SAR-CoV-2 spike protein.

necessarily use exactly the same sequence that is in the virus. We want to make a gene that will produce exactly the same protein sequence, but will do that really efficiently in human cells. That means using a technique called codon optimisation. Sarah S was going to use that and other techniques to design the exact DNA sequence we needed.

The four letters in the genetic code (A, C, G and T) provide the instructions to make all of the proteins in our body, and proteins are what ultimately enable every form and function of our body, from the shape of our muscles to the digestion of our dinner (using enzymes, which are one form of protein). Proteins are made up of combinations of twenty different amino acids. Think of it as like making necklaces from different colours and sizes of beads. If you have twenty different types of bead (amino acid), you can make lots of different kinds of necklaces (proteins). The genetic code, or DNA, provides the instructions to put amino acids in order.

For example, one of the amino acids is called methionine. Every protein starts with methionine, although it can also occur in other places in a protein. The code in the DNA that says 'put methionine here' is ATG. That group of three letters is called a codon. If the next codon is TGG, that means the next amino acid will be tryptophan. For all the other eighteen amino acids, there are multiple codons that lead to the same amino acid being included. For lysine, the code is either AAA, or AAG. For arginine, it can be CGT, CGC, CGA, CGG, AGA or AGG. Different organisms prefer to use some codons rather than others. For example, the bacteria that cause tuberculosis have a very 'GC-rich' genome (meaning they use codons that contain a lot of Gs and Cs), whereas malaria parasites are very 'AT-rich' (so they use codons that contain a lot of As and Ts). TB bacteria and malaria parasites can both encode all of the amino acids, but they are more likely to use some codons rather than others. In humans arginine is more likely to be AGG than CGT.

If we want the new protein we are coding for to be produced in large quantities inside human cells – which we do, because that will make the vaccine more efficient at stimulating an immune response – we should use the codons that are most common in humans. So, once Tess had identified the part of the genetic code we wanted – 1,273 codons, made up of 3,819 letters of genetic code, providing the code for the spike protein – Sarah S, using specially designed computer software, went through the process of changing the DNA sequence to use the common human codons rather than the rare ones.

She also joined a short extra sequence onto the beginning of the spike protein's code. This was another technique that we had developed to make the vaccine more efficient. The extra sequence coded for thirty-two amino acids that are found at the start of a human protein called tissue plasminogen activator. Adding that sequence had been shown in other vaccines to lead to a stronger immune response, because it first tells the vaccinated human cell to make larger quantities of the spike protein, and then cuts itself off from the protein once the protein has been produced.

Within forty-eight hours of the novel coronavirus genome being released, Tess and Sarah S had chosen the exact protein sequence they wanted to encode and the precise DNA sequence that they needed for that. The order to get this new piece of DNA made was sent off to ThermoFisher, a kind of life-sciences supermarket who sell everything, including custom-made synthetic DNA. (The order would have looked like a roughly 4,000-letter string of different combinations of the letters A, T, C and G.) And then it was back to the discussions over exactly what to do with it when it arrived.

On that Saturday, 11 January, there was still no clear evidence that human-to-human transmission was taking place. There had been no confirmed cases outside China although there had been one suspected case in South Korea. And I was still sifting DPhil applications and telling my friends not to worry.

On the evening of Monday 20 January, I received an email from Sarah G asking if we could meet to 'discuss a couple of things'. Over the weekend, the number of confirmed cases in China had risen from 41 to 201 and earlier that day China had confirmed human-to-human transmission was taking place. Sarah suggested that this news made it more likely that her vaccine project would move from experiments in her lab to demonstrate what was theoretically possible, to actually making a vaccine for human trials. Meaning, I was going to need to get involved. I replied the next day (I am not a same-day email replier, an aspect of my personality that has been severely tested this year: by March, if Sarah emailed me I would be replying within twenty minutes) and we set a time to meet. It still didn't feel particularly urgent.

A couple of days later I headed across the research campus to the Jenner Institute where Sarah is based. As I had done many times before, I passed through the airy atrium where students and scientists were gathered in small groups drinking coffee or in intense discussion over a laptop. I climbed the concrete stairs two floors, and sat down in Sarah's little office, full of university-issue furniture and box files. Sarah's familiar 'keep calm and make vaccines' mug was on her desk. It seems strange looking back on it now, but the meeting was relaxed. Neither of us, I think, had yet registered how much of an impact this virus, and this work, was going to have on our lives.

Sarah set out the situation. She and Tess had already ordered the DNA sequence, anticipating that they might get funding just to do some testing in the lab and animal experiments. Now she wanted to know, would the CBF be able to start making the starting materials for a vaccine against this new virus, immediately, using the rapid method?

I wasn't surprised. My colleagues are always coming to me asking me to do a lot of work, by tomorrow, for no money. My answer was measured. Of course, we *could* do that. The pre-GMP lab, where we make the starting materials, theoretically has capacity to work on four projects at any one time, although we didn't actually have enough staff to do that. And I could see that this was potentially important. But the team were all currently working on other projects: one for Sarah (a vaccine against Crimean-Congo haemorrhagic fever) and one for Tess (against Lassa and Marburg viruses), so prioritisation of a vaccine against this new coronavirus meant these would slip.

My other concern was financial. The CBF is run like a small business within the university, and in order to operate it has to cover its costs of around £1.5 million a year by charging its clients – researchers like Sarah and Tess – who in turn have to apply for grants to fund their research. The projects that Sarah was asking me to delay or deprioritise were already agreed, and their funding was secure right through to manufacture. It would be a big risk to the CBF's operation to drop those in favour of this new project, when it wasn't at all clear where the money might come from to pay for any of it.

Sitting in Sarah's office, I had to balance the risk to CBF finances of losing already-agreed work, with the fact that not going as fast as possible would be something to regret if the spread of the virus continued to escalate.

'OK,' I said. 'Let's go for it.' I was in.

The CBF would make the starting materials for the vaccine – about 100 ml of 'seed stock' from which all batches of clinical-grade vaccine would be made. Sarah would approach a selection of manufacturers we had worked with before to try to secure manufacturing slots to make the clinical-grade vaccine later in the year, and also, as a backup, feel out the funders of our next clinical-grade manufacturing project, due to start later that year, about the possibility of delaying if necessary. I headed straight back across campus to the CBF to begin to work up a plan and let my team know what was going on.

Over the next few days, Wuhan locked down and started building hospitals, and cases were reported in ones and twos in countries across the world. As we waited for the synthetic DNA we had ordered to be shipped back to us, Sarah, Sarah S, the CBF team and I discussed which method we should use to make the starting materials: the still-not-fully-developed rapid method, or the tried-and-tested classic method. We agreed to try the rapid method – because of course we wanted to move as quickly as possible: this no longer felt like mostly an intellectual exercise – but also to initiate the classic method separately as a backup.

By this point in late January Sarah had found some funding for us to cautiously start making our own high-quality batch of the ChAdOx1 vaccine needed to kick off the rapid method (baking the cake in advance). We were about halfway through and would take another three weeks or so to complete that process. On the Friday of that week, that was still our plan.

On Saturday 25 January, the synthetic DNA was shipped back to us. Inside a courier-delivered Jiffy bag was what looked like an empty test tube. In fact it contained about 100 billion dehydrated strands of the DNA sequence coding for the spike protein. This was invisible even under a microscope, and weighed a few micrograms (far less than a grain of sand, which weighs a few milligrams).

By the Monday, our plan to 'bake in advance' our own ChAdOx1 vaccine, ready to be 'decorated' with the gene for the spike protein, was coming under pressure. We had the DNA we needed to start decorating, everything else was lined up, and the clinical situation was looking increasingly worrying. It no longer felt right to wait three weeks while we pre-baked our cakes. Instead, Sarah S gave us some that Vaccitech had prepared earlier. They had also been working on the rapid method in preparation for using it to make personalised cancer vaccines. And so we embarked on our first ever attempt to make a vaccine using the rapid method.

By now it was 27 January. We had designed the vaccine, agreed that the CBF would produce the starting material, decided to progress with both classic and rapid methods in parallel, and begun all the necessary paperwork to show that we could justify our decisions and that risks had been considered and where necessary mitigated. Sarah had found some funding that would tide us over for a while. My team at the CBF were prepared and ready to get started. And we were already considering ways to speed up other slow parts of the manufacturing and release process.

It was also at around this point that Sue Morris, an adenovirus expert who works at the Jenner Institute with Sarah, sent an email saying she was losing sleep. Sarah had come up with the original idea for the rapid method and how it would work in principle, but it was Sue who had done the lab work to make it happen. Now, Sue was kicking off the classic method in Sarah's lab, and at the same time holding our hands through the rapid method over at the CBF, because she knew more about it than anyone else. This would be the first of many similar emails: sleepless nights were to become the new normal, but at this stage we were all naively thinking that things would calm down once we got going.

—

I had a few days in Paris booked at the end of the month for my birthday. I had lived there for three years from 2000, working in a laboratory at the Institut Curie funded by an EU grant. My old friend Helen and I drank cocktails in crowded bars along the canal Saint-Martin and I wandered my old haunts, feasted on charcuterie and Brouilly squashed in at tiny tables in my favourite cafés, and along with a horde of other tourists saw the Leonardo exhibition at the Louvre.

Some of it felt just like in the old days, and yet, not quite. My friend Leo has an apartment in the Belleville area, known for its Chinese restaurants. Three cases of the new virus had been confirmed in France on 24 January, all people recently returned from Wuhan. Leo said that already the neighbourhood felt quieter, more fearful, as people realised that this was not just a problem that would affect someone else, but that this virus was now spreading across the globe.

I returned from Paris by Eurostar late on the evening of Friday 31 January, my bag full of cheese and wine. I was on nearly the last train to London while the UK was still a part of the European Union. It felt like the end of an era, and the atmosphere on the train was subdued.

We didn't know then that while it was indeed the end of that era, we were right at the start of a new one that none of us had voted for. That day, in York, the UK recorded its first confirmed cases of the novel coronavirus. My memories of this trip to Paris would soon feel like those from another, radically different, time.

CHAPTER 4

Money, Money, Money

13 January–21 April 2020[1]
Confirmed cases: 60–2.57 million
Confirmed deaths: 1–183,458

'Where's the money coming from?' Cath asked. It had been a tense few days and this was a reasonable question. At this point there was no money.

'We are the ones who can do this,' I said. 'We're just going to have to do it anyway and work out the money later.'

That was early March. From January through to April, we and the world came to understand the enormity of what was happening, and our lives changed in previously unimaginable ways. In January, the unfolding fate of the residents of Wuhan was sometimes making it into the news headlines, accompanied by reassuring words about the UK's state of preparedness. By the end of March, we were locked down and clapping for carers, there had been more than 875,000 confirmed Covid cases and more than 44,000 confirmed Covid deaths worldwide, hospitals and care homes were scrabbling for PPE, and our prime minister was self-isolating at Number 10. It would be an understatement to say things escalated fast.

In January, Tess and I were interested in how quickly we might be able to make a vaccine against a brand-new pathogen – understanding that it might turn out to be nothing more than an intellectual exercise. By the middle of April, the whole world wanted to know.

I spent a huge amount of time during those increasingly strange weeks attempting to secure the funding we needed. At this point, money was my main preoccupation as it was the most likely reason that we would fail to make a vaccine. Thankfully, the worst of this very stressful enterprise was over by the end of April: by that point we had over £22 million in committed funding, and a commercial partner. But in those early months, I thought of little else, day and night. We couldn't wait for research funding to be awarded – which you will understand, having seen in previous chapters how painfully slow and uncertain this process can be – so we decided that we just had to get on with it, spending money we did not yet – and might not ever – have. We would ask for forgiveness, not permission.

I wasn't sleeping a great deal through that time and I was juggling a lot of balls. Much of it is a blur. But I do remember that on Saturday 8 February, when I began writing the application to the governmental body UK Research and Innovation (UKRI) for just over £2 million to fund the manufacture of our vaccine for our first clinical trials, I was radioactive. I don't mean I was 'on fire' or 'on top of my game'. I mean I was literally radioactive. I was undergoing a diagnostic procedure that involved swallowing a capsule of radiolabelled bile acids.

I travel a lot as part of my work, or at least I did until the end of 2019. Because my focus has been working on vaccines against diseases that are prevalent in low- and middle-income countries, and because cutting-edge scientific research is by nature very international, I've taken work trips to North, South,

East and West Africa, China, Saudi Arabia, the United States, much of South East Asia, and quite a bit of Europe. Often trips last only a few days: a two-day trip to Delhi, economy travel both ways, and only leaving the hotel to be driven back to the airport, was not unusual. For the last several years I had increasingly been suffering on these trips from 'traveller's tummy'. I knew all the advice about what foods to avoid, but I am an annoyingly fussy eater and sometimes found that advice difficult to follow.

During the summer of 2019, I decided to google my symptoms and came across some information on a condition called bile acid malabsorption. Apparently, it often goes undiagnosed and tends to get worse as people get older. When we eat food containing fat, our gall bladder releases bile acids into the stomach to break down the fat and allow it to be digested. The bile acids pass out of the stomach along with the food and are reabsorbed in the small intestine and recycled. However, in some people the reabsorption doesn't happen properly. That means the bile acids carry on through the digestive tract to regions they are not supposed to be in, causing irritation and leaving the owner of the digestive tract in question needing to rush to the loo.

This seemed a plausible explanation for my issues. My GP concurred, but on the other hand I might have bowel cancer and she would be remiss not to rule that out. So she referred me to a consultant gastroenterologist for tests. The consultant turned out to be a very pleasant woman who told me that she knew of my work because her former PhD supervisor was a collaborator of mine. (This kind of thing is an occupational hazard. Oxford is not a large city, but there is a huge amount of medical research going on, and a lot of hospitals. When I went into hospital, in labour with triplets at thirty weeks' gestation, the consultant who came to see me was someone for whom I had previously organised a genetics seminar. If it hadn't been him, it could have been the other obstetrics consultant, who was in my Pilates class.)

This particular consultant arranged a series of tests for me, which all came back as normal. By the end of the year, the only test still outstanding was a so-called SeHCAT test. For this, a capsule of radiolabelled synthetic bile acid is swallowed, and three hours later a scanner is used to detect how much radioactivity (and therefore bile acid) has been absorbed into the gall bladder. A week later another scan is taken to see how much remains. If the answer is 'very little' it means that the bile acids are not being reabsorbed in the normal way.

So that's why, on Saturday 8 February, my insides were literally radioactive and my mind completely occupied with thoughts about funding. Cases of the (still unnamed) novel coronavirus were by this time starting to be reported in dozens of countries across Asia, North America and Europe, including three confirmed cases in the UK. I had been able to get early work on our vaccine off the ground using a small amount of flexible funding from a project called VaxHub. But to get any further, I urgently needed to secure much more.

Actually, raising funds had been my main activity for years. When people think of scientists, I think they imagine us bent over complex equipment in a lab, or staring intently at a test tube. That used to be me and I often wish it still was. I enjoyed the many small triumphs every week in the lab – creating a perfect image of the DNA I had just produced and cut up, or noting how the numbers came out exactly as I had expected on a virus titration.* I trained for years to become really good

* A titration is a common lab method for determining the concentration of something – in this case, a virus – in a sample.

at 'doing science', and I was good, too, at training other people to be good at it. But by early 2020 I hadn't worked in a lab for over ten years.

What I actually spend my time doing these days is, mostly, bringing in the money. I am employed by Oxford University to conduct research rather than to teach undergraduates. This means that I am required to secure the money to fund that research. I have to find ways to cover the costs of the equipment and materials we use in the lab, like tissue culture vessels or growth medium for culturing cells. I have to cover an overheads charge to pay for the running costs of the university buildings we work in, and so on. I also have to cover my own salary and the salaries of the people who work in my team: mainly clinicians and lab researchers, but also three full-time project managers and a contract specialist, to help secure the funding, keep track of it, and report back to our funders. Essentially, each research group is like a small business or charity. It is made up of a team of people working on a set of projects, and the head of the research group (also known as the principal investigator) has to bring in the funds to keep everyone in the team employed. This causes immense stress and frustration for researchers, and can be counterproductive for the cause of scientific research itself.

For example, sometimes grants are awarded to fund research into a particular area, with some flexibility as to how the funds can be spent. Increasingly, though, funding is for a quite specific purpose. And sometimes we apply for and are awarded contracts rather than grants. Contracts are more prescriptive than even the most restrictive grant: they sign us up to do an exact project, in this exact amount of time, for this exact amount of money. They can be appropriate and work well if we can thoroughly plan out everything we are going to need to do in advance

– for example, if we are manufacturing a vaccine for a clinical trial and then conducting that trial.

However, the lack of flexibility in a contract can be problematic. For reasons connected to government funding cycles, we have to agree to start the work by a certain date, report every three months, and then finish by a certain date. This means researchers are often trying to shoehorn the work they want to do into the time and budget available, while project managers spend a lot of time tracking the researchers' activity and filling out more forms applying for more time, or more money, or some other variation to the contract. Someone in the team taking maternity leave or moving to another post can throw a whole project off course. More importantly, the lack of flexibility leaves no room for creativity, or for discovery. The strict structure expects us to know what we are going to do and how long it will take to achieve it – which works when the research can be defined in advance but not when it is exploratory and innovative.

Obtaining funding in normal times is also a long and uncertain exercise. The applications are complex to write, the process frequently takes a year from start to finish, and the chances of success are always well under one in three (as you'll have seen with our failed CEPI application in Chapter 2). Research councils and other funders – in my area, that essentially means CEPI, the Wellcome Trust, UKRI, and until recently the EU – will put out 'calls for proposals', inviting applications for work in a particular area to be submitted by a set deadline. Of course, each funder has its own application process, with its own quirks, involving different ways of setting out what work is to be done, what it will cost, and how long it will take. Each funder also has different rules about what it will pay for. For example, the Wellcome Trust won't pay for any PPE. In research labs, researchers need disposable gloves – this is in part to protect the people doing the work, but it's also to protect the work from us, as humans have bacteria and enzymes on our skin that could corrupt whatever we're working on. Safety glasses and lab coats are also mandatory in most labs. On the other hand, CEPI will pay for PPE, but won't pay for stationery.

People may think of science careers as being very stable. But in reality working in academic research can be extremely uncertain. A grant might provide funding for a project lasting anything from a year to five years. But the university only allows us to advertise jobs for that project once the money has been awarded. This means more waiting. It might be months before someone actually starts work on the project, but their employment contract can't extend beyond the period covered by the funding, which means someone employed to work on a three-year project who doesn't arrive until six months in will get a contract for two years and six months.

As they near the end of their contract, researchers often don't know whether their principal investigator will obtain further funding so that their work, and employment, can continue. It is stressful and researchers have to choose between devoting their efforts to getting the research done and papers published – thus increasing the chances of more funding – or looking around for other jobs. Many excellent scientists, Sarah Sebastian being just one example, leave academic research because of the lack of stability and the continuous pressure.

So, given how much time has to be spent securing funding, and writing papers, and attending meetings all over the world to hear about the latest developments (and raise one's profile), it is rare for a principal investigator like me to continue with any lab work at all. At various points in the decade prior to the pandemic I regretted this. But in early 2020, I did not. Instead, I looked on those years of navigating the complexities and frustrations of endless grant applications as vital training. Without having built up that experience, I would never have been able to make my way through the fiendish funding maze that sprang up in the first part of 2020, and get our vaccine project so far so fast.

—

In the early days, when we needed to get the project off the ground and the world wasn't yet alive to the urgency of the situation, I was able to get going with funding from VaxHub. All of the other funding I already had was for very specific projects. But the VaxHub funding was for vaccine development work in general and that allowed us to make a start.

VaxHub, based at University College London, was a three-year programme that was aimed at improving vaccine manufacturing methods and engaging with both UK and overseas manufacturers – specifically, manufacturers from low- and middle-income countries. It was important because manufacturing can be a big bottleneck in getting new vaccines out there and into people's arms: it's no use making a fantastic new vaccine that looks great in the lab, but that can't be manufactured in a way that allows it to move into clinical testing. Going back and reinventing it to make it suitable for manufacturing can add years to the development process. The funding for VaxHub came from the UK government's Official Development Assistance budget, meaning money set aside to invest in development in other countries. The UK had committed to spending 0.7% of its gross national income on this in 2015, and had recognised, following the 2014 Ebola outbreak, that it could put some of that money to good use on work to develop vaccines against diseases that caused havoc around the globe.

The funding from VaxHub got us through January. It allowed us to buy the synthetic DNA we needed, make the first batch of vaccine in the lab, immunise some mice and analyse the results. But even as Tess and I were getting started with designing the vaccine on 11 January I knew that if we wanted to keep going, we needed much more.

Vaccine development funding can roughly be divided into four phases: design and preclinical testing; making and testing the starting materials; making the vaccine for, and running, the clinical trials; and finally large-scale manufacturing and roll-out. And it gets more expensive the further you get – each phase requires at least one more zero on the end of the costing. Our main prospects were CEPI and UKRI, so I started scouring their websites and emailing my contacts there.

On 13 January I initiated discussions with CEPI. They said that they were going to 'focus on the very quickest platforms – namely DNA and mRNA' in the first instance. I was astonished. DNA and mRNA vaccines can be ready for clinical trials quickly, but not necessarily much faster than adenoviral-vectored vaccines. (In the event our phase III trial was the first to start by a long way.) This could certainly be the chance to use mRNA vaccines on a large scale for the first time, but committing to only DNA and mRNA vaccines made no sense to me. It was a disheartening start.

The WHO declared a public health emergency on 30 January. This is the highest alert level that the WHO can announce so even though the word pandemic was not used, the world should have taken more notice of it than it did.* The next day the first

* A public health emergency of international concern (PHEIC) is the highest level of alert that the WHO is obliged to declare under the International Health Regulations. A PHEIC is defined as an 'extraordinary event that constitutes a public health risk through the international spread of disease and potentially requires a coordinated

UK case was confirmed. Despite what CEPI had said in January about focusing on mRNA and DNA vaccines, they put out a more general call for 'proven vaccine technologies, applicable for large-scale manufacturing, for rapid response against novel coronavirus, 2019-nCoV'.[2] The closing date was two weeks later, which was extremely tight. Not long after, UKRI put out a similarly wide call for proposals for almost any research that might help to tackle Covid, with £20 million on offer altogether.[3]

Part of successfully winning funding in these competitive situations is knowing how much of the total pot to bid for. Assuming I got the CEPI funding to make and test the starting material, the next thing I would need was funding to manufacture the first batch of clinical-grade vaccine for clinical trials, and to set up and then run the clinical trials themselves. Neither of those activities come cheap, but I knew that if I asked for much over 10% of the total pot I was unlikely to get it.

Fortunately, because of the work we had done in January, we had already produced a batch of research-grade vaccine, immunised some mice with it, and shown that it produced an immune response. Our vaccine genuinely was ready to take to the next stage, and I felt our application was strong. But time was short and the stakes were high and everything had to be lined up.

One part of the puzzle was manufacturing. Cath's only 'clean room' was already taken, manufacturing an Ebola vaccine for

international response'. The purpose of such a declaration is to signal the need for urgent international action and oblige countries to share information. Previous PHEICs include the H1N1 flu outbreak in 2009 and the Ebola outbreak in 2014. On 11 March 2020 the WHO additionally characterised the spread of Covid-19 as a pandemic. 'Pandemic', meaning an epidemic affecting many countries, is not an official WHO alert level.

Tess, and it didn't feel right to pull that work.* ('Clean room' is the name for the strictly regulated lab in which we are able to make vaccines for human use.) But our long-time collaborators at Advent in Italy, a contract manufacturing organisation specialising in making adenovirus vaccines, had availability in April and May, so I asked them to quote for doing the manufacturing with the starting material Cath and her team were now making.

The plans for the clinical trial itself had to be quite modest, and even then I knew they would need to be expanded later, but this at least would get us started. Coronaviruses are respiratory viruses. Knowing that with respiratory viruses children often become infected and transmit the virus despite not being noticeably ill themselves, I wanted to include immunisation of children in the clinical trials. The Oxford Vaccine Group, part of the university's Paediatrics Department specialising in running vaccine trials in children and headed up by my colleague Andy Pollard, had enormous experience and so working with them made much more sense than trying to set this up from scratch within my own group. As it turned out, children don't play such an important role in transmission with Covid-19, and we learned quite quickly that having a vaccine that was effective in older people would be far more crucial. Because of that we decided to wait for the vaccine to be approved for use, demonstrating that it was highly safe and effective in adults, before enrolling children into the trials. The children's trials were eventually started in March 2021. However, asking Andy to join the team turned out to be a really smart move anyway – not only because

* You may be wondering why Tess was making an Ebola vaccine in 2020 when you know from Chapter 2 that an effective VSV-vectored vaccine had been developed back in 2014. There are two species of Ebolavirus known as Zaire and Sudan. Tess was trying to make a vaccine that would be effective against both variants.

of his huge experience but also because of his unwavering commitment to public health and capacity for doing whatever it takes to make a difference.

So this was the state of things when on Saturday 8 February, I walked to the John Radcliffe hospital in Oxford to swallow a radioactive capsule. I was due to be flying to Amsterdam the next day for a meeting about the MERS, Nipah, Lassa vaccines programme, so I was also given a letter to take with me to the airport the next day, explaining why I might set off the security alarms. Looking back now it seems slightly surprising that, not only had I not dropped my other work at this point, I also was prepared to get on a plane to Amsterdam. (In the event, the meeting was cancelled because my colleagues at Janssen were also very busy with their own Covid response work.)

Having taken the capsule, with instructions to return in three hours, I walked back to my office to work on the grant application. Filling in these forms can be a bit like taking an exam – there are strict word limits so you have to consider carefully what is being asked in each section and answer precisely, with no waffling. I have honed my technique over time, but it's still about as fun as taking an exam. As the radiolabelled bile acids were making their way to my gall bladder, I worked my way through the application form.

A week later, I was back in the hospital looking at an image of my abdomen. Lying still on a hard surface with a pillow thoughtfully provided and the scanner over my abdomen, I was the most relaxed I had felt for a while. The background was black, and there was a faint, roughly circular patch of white dots, which apparently was my gall bladder, with a few more white dots scattered over the larger area of my abdomen.

Eventually, about eight months after first suspecting that I was suffering from bile acid malabsorption, I got a letter confirming it. This did not come as any great surprise. I had been pretty confident eight months previously that this was what was going on. The GP had been pretty confident too. And the simple treatment that I had started taking anyway had already been working. But I had to do the tests to make sure, and to rule out something worse like bowel cancer, which inevitably took time. I'm telling this story as it strikes me that there are parallels between getting this kind of official diagnosis, and vaccine development. Similarly, I was already pretty confident in January that we had designed a vaccine against the novel coronavirus that would be safe and effective. But we couldn't just assume that. My hypothesis would have to be tested, and worst-case scenarios – like a vaccine that was not safe – ruled out. And again that would, unfortunately, take time. How much time would depend to a large extent on the money.

—

The applications to CEPI and UKRI were both submitted in mid-February, but I knew it would be several weeks before we heard back from either of them.

Meanwhile, it was becoming clearer to me every day, as the virus continued to spread across the world at unprecedented speed, that we could not afford to lose time waiting. And even when the money did come through – though really it was an 'if' rather than a 'when' – it would still barely touch the sides of what we needed.

In normal times, we would have an idea for a new project, write a grant application, and only when we heard that we had the funding – probably about a year later – would we order the gene and start the work. But mid-February 2020 was not a normal time. We had ordered the gene on 11 January and it had arrived back with us two weeks later. Cath and her team at the CBF had begun making the starting material. We had clinical trials designed and, working with Andy and his team, needed to start preparing for them. Together with another colleague, Dr Sandy Douglas, we were already thinking about large-scale manufacturing. Cath and I had started talking to the regulators. In any other circumstances we would have waited, before doing any of it, for funding to be confirmed. Instead, through February, into March, and for the first half of April, we had to keep going, cross our fingers and hope that the money would come through.

Some of those decisions we could take on our own: in a university, academics can act relatively autonomously, at least for a short while, and Cath and I could take a deep breath, brace ourselves, and just decide that she was going to commit the time at her very expensive facility. For other things we did need permission, though.

For example, I had included in the UKRI application the cost of manufacturing the vaccine at Advent in Italy. But to reserve the precious time in Advent's clean room, there needed to be a contract. Usually it would simply be unthinkable to ask the university to sign a contract before we had official confirmation of the funding. However, this was a pandemic. Without the contract, we would lose the manufacturing slot we had lined up at Advent. Without the slot, our clinical trials would be delayed. There had to be a way. Going against normal protocol, the university agreed to take the risk and underwrite the contract. If no funding came through, the university would cover the costs.

Forging ahead despite no guaranteed funding solved one problem – it meant we weren't waiting. But it wasn't enough. In the race between us and the virus, the virus was drawing ahead. To have any chance of keeping pace with it, we needed to work harder and faster, and that would require more funds than I had felt able to bid for from CEPI and the UKRI pot combined.

The obvious place to look was CEPI again. I called them up and asked if we could use some of the $19 million that my lab had been awarded in principle for our MERS, Nipah, Lassa project. That award included funding to work on a large-scale manufacturing process for ChAdOx1-vectored vaccines. Since this work would only need to be done once, and would be applicable to all ChAdOx1-vectored vaccines, could we spend the funds now, for work on large-scale manufacturing of the Covid-19 vaccine? I spent a lot of time discussing this with senior people at CEPI during February and March, and writing an application to do it. One difficulty was that funders want a definite plan – what exactly are we going to do, how long will it take and what will it cost. But our plans kept changing, getting bigger and faster every day as it became increasingly clear how urgently the world was going to need this vaccine. For example, there came a point in early March when Cath and I realised that waiting for Advent to manufacture the vaccine would take too long. Cath and her team would need to stop work on the Ebola vaccine and switch to manufacturing the first batches of our vaccine, snappily known as ChAdOx1 nCoV-19. We didn't have the funding to pay for the manufacture at the CBF, but we all agreed: 'We're just going to have to do it anyway.'

We also started lobbying the government through our connections at the UK BioIndustry Association (the BIA, a network of companies that work to manufacture or support the manufacture of biological medicines). We were contacting anyone we could find to say 'We think we can make a vaccine, and we think we can make it fast, but we are going to need money.

This is not a cheap process.' And we started looking for a commercial pharmaceuticals partner who would be able to manufacture our vaccine at the large scale that it was now obvious would be necessary.

Just five weeks after CEPI had put out a call for proposals, we got the funding. We now had $350,000 for making and testing the starting material. Five weeks: compare this with the twenty-one months it had taken to get from proposals to funding for our MERS, Nipah, Lassa project. It provided some small relief. But the reality was that we were already halfway through the work this money was meant to fund, and were doing much more that continued to be unfunded.

Two days later, on 10 March, I took the train to London. I had been invited to speak about our work on a possible Covid vaccine at a private briefing of the government's Science and Technology Committee and attendance in person was still, just about, what was expected. The last time I had been to the House of Commons had been a few years earlier when our whole family went on a tour as part of a day out in London. I took a train to Marylebone station and walked from there, through Marylebone and Mayfair and across St James's Park. I didn't want to go on the Tube if I could avoid it, because of the risk of being crammed together with people who might be infectious, so I enjoyed the walk through the unusually quiet streets.

I was one of six experts invited to give evidence to the committee. Assembled beforehand outside the committee room, we all kept our distance from one another. Then one of the other experts who I had not met before held out his hand to shake mine. I responded with a Lao-style greeting, placing both my hands together in a prayer position and giving a small nod. Yes, we shouldn't be shaking hands, maybe we could bump elbows instead, he suggested somewhat uncertainly. I said that I'd rather use a greeting that involved no contact at all. (To be honest, handshakes had caused me problems in the past and I would not be sorry if they were a permanent casualty of this pandemic. On a visit to Riyadh to discuss a MERS vaccine clinical trial, I knew that men might not want to shake hands with a woman, and so didn't offer my hand. At a conference I was to be invited onto the stage to receive an award from the guest of honour, a Saudi prince, and was told, 'On no account attempt to shake hands with the prince.' But as I walked towards him, he held out his hand to me, so I shook it. By the end of the day, having reverted to the practice of offering my hand, a man recoiled from me in horror, his hands firmly behind his back, saying, 'I don't shake hands with strangers.' By which of course he meant women. Establishing a new universally recognised greeting with no physical contact could solve a number of problems.)

In the committee room in the Palace of Westminster I spoke about the vaccine development work we were doing in Oxford, how it followed on from other work we had done on the MERS vaccine, and why it was both possible and essential to proceed very quickly. I also said that we needed funding for the work as a matter of urgency. The following day, 11 March, the WHO declared Covid-19 a global pandemic.

The next few weeks were some of the most hectic and surreal of my life. I was running in parallel half a dozen stages of vaccine development that would usually happen over years and in sequence. At the same time we were starting to attract a lot of media interest, setting up a deal with big pharma, and going into lockdown. But by the end of those weeks, while I was concerned about many other things, our money worries seemed to be over. Finally it felt like we had the support to be able to do what we needed to. At the end of March, just as the entire country went into lockdown for the first time, we heard that my radioactive bid for £2 million from UKRI had been successful. Which was lucky, because we had already spent quite a lot of it. Cath and her team were several weeks into making the vaccine whose manufacture the money was supposed to fund; and of course that contract with Advent had been signed.

On 17 April the UK's new Vaccines Taskforce was announced. Its role was to support efforts to develop, manufacture and roll out a vaccine as soon as possible, including by 'rapidly mobilising funding', and on 21 April the government awarded us £20 million (later expanded to £31 million) to support both expanded manufacturing and clinical trials. Again we had already started spending it: our clinical trials started two days later. We also were not guaranteed the whole lot: £10 million was awarded initially, but we were going to have to jump through a hoop partway through the project to release the second half. Andy and his team needed these funds to get started on expanding the clinical trial from Oxford to eighteen other sites around the UK, so I had to come up with a list of milestones that would demonstrate significant progress had been made but that I was confident we would achieve very soon. By early May we had secured the second half of the £20 million, which meant the phase III trials could begin.

Usually, securing $350,000 from CEPI,* or £2 million from UKRI, let alone £20 million from the Department of Health, would be the cue to throw parties of ever-increasing size for everyone involved, which was a continually growing group of people. But of course, we were locked down. And we still had

* Later on, CEPI provided up to $383 million for AstraZeneca to manufacture 300 million doses of the vaccine for Covax, the global facility committed to securing vaccines for low- and middle-income countries.

a lot to do. There was no time to celebrate, and no possibility of coming within two metres of one another, let alone throwing a party. It didn't even, really, feel like a success. It felt like a disaster averted. It would have been completely awful if we had not got the funding: our work would have stopped, and our names would have been mud for the next ten years as the university paid off our debt. So there was no party, just huge relief all round. Now that we had finally solved the money, we could get on with the science. I celebrated by going to bed ten minutes early.

CHAPTER 5

Making the Vaccine

28 January–22 April 2020
Confirmed cases: 5,578–2.65 million
Confirmed deaths: 131–190,322

We were never searching for, hunting for, or, my personal bugbear, finding a vaccine. All year, the papers talked about 'finding' a vaccine, as though the thing already existed, maybe in a forgotten freezer somewhere, and our only job was to look hard enough. We never talk that way. We are not some kind of Lara Croft in a lab coat, out in search of hidden treasure. Vaccines are not found. Even the world's first vaccine, against smallpox, was the result of a carefully thought-out line of reasoning, not a happy accident.

Vaccines are first designed, based on scientific knowledge accumulated over decades and detailed study of the pathogen. Then they are made, carefully and painstakingly, by teams of people following strict rules and regulations. And then they are tested thoroughly for safety and efficacy, again according to strict rules and regulations.

This chapter is about how we first made our Covid-19 vaccine. The techniques we used range from tried-and-tested methods that a sixteen-year-old science student might recognise, to newly developed, state-of-the-art technology. The work was fiddly and exacting. The people, often manipulating tiny quantities of precious substances and working under layers of protective gear, were meticulous, systematic and dedicated. As we've been explaining in this book, we were not completely unprepared: we had been thinking about, and trying to prepare for, Disease X for some time. But one thing we had not thought about was this: how do you fight a pandemic, when you are in a pandemic?

—

At the end of January, before I left for my birthday trip to Paris, I sent an email to the whole team at the CBF. The email let them know that we were immediately beginning work on the starting material for a vaccine against the coronavirus from Wuhan that they had heard me talking about over the last few days. At this stage that was all we were committed to do. However, I was clear to note, 'This is an evolving situation and so our plans may change on an almost daily basis if the situation globally becomes more or less severe.'

Different kinds of vaccines are made in different ways. When it comes to making any adenovirus-vectored vaccine, which is what our Covid-19 vaccine is, I always tell people it is a bit like making bread. Partly because making vaccine from starting material – which is unique to adenovirus-vectored vaccines – is like making sourdough from a starter. And maybe also because during the first lockdown baking became quite a focus for me as a way to stay sane. My family's weekly Zoom baking sessions, with my parents, my sister and her family, and me and my daughter all cooking the same recipe and then comparing our creations – everything from Cornish pasties to chocolate Swiss roll – really helped to get me through.

There are five steps in the recipe for any adenovirus-vectored vaccine:

Step 1: Make the starting material. A vaccine's starting material is like a sourdough starter – it acts as the basis for all the batches that will eventually be made.

Step 2: In a clean room, use the starting material to produce the required quantity of vaccine. In baking terms, this is when you mix the ingredients and bake the bread.

Step 3: Still in the clean room, purify the vaccine away from unwanted cellular components used in the manufacturing process. (Stretching the analogy a bit, this is like removing the bread – the bit you want – from the tin – the bit that was useful for making it, but that you don't need any more.)

Step 4: In a completely sterile environment, fill into individual vials. (Slice and wrap your baked bread.)

Step 5: Label the vials, document the process, test and certify the product, and distribute. Essentially – the taste test before putting on the shelves.

In this chapter, using these same steps, I'm going to walk you through how we first made (not found) the vaccine that is now being rolled out across the world and may now be protecting you and your family from this horrible disease.

Step 1: Make the starting material

We normally allow three to four months to make a starting material in the CBF using the classic method, but this time we were going to be attempting the rapid method. At this point, my most optimistic estimate was that we might have a starting material ready by mid-March, and the vaccine (made by Advent) ready for clinical trials (step 5) by the end of July. To hit these dates everything would need to go perfectly. The rapid method, which we had never used before, would need to work. The team would need to do double shifts and come in on weekends. We would need to deprioritise all other projects. And we would need to proceed through the process 'at risk' – in other words moving forward to the next stage of the process before all testing was complete on the previous stage. It's important to be clear here that the risk is not to the end product, the vaccine itself, but to our time, effort and finances: if a problem is identified during testing, some of our work has been futile and we have to back up and repeat a step. In the event, many things did go wrong. But our first batch of starting material was ready, as we had hoped, by 17 March and the vaccine was ready for trials by 22 April, much earlier than we had originally hoped.

On Tuesday 28 January, after the first of her many sleepless nights, Sue and the pre-GMP team kicked off the rapid method. The pre-GMP team make the starting material. Sue took the pre-baked ChAdOx1 vaccine that Vaccitech had prepared earlier, and combined it with the synthetic DNA that coded for the spike protein. Now she, hopefully, had a sequence of DNA that coded for the Covid-19 vaccine, ChAdOx1 nCoV-19. But virus DNA outside of a living cell is just an inert chemical. To allow the virus to assemble and replicate itself, the DNA needed to be inserted into living cells, which would become little microfactories, churning out copies of the virus. Sue and the team introduced the DNA into a special type of lab-grown human cell known as HEK293 cells.*

The method we use for inserting naked DNA into a human cell is called transfection and has been used for decades. The external membrane of human cells is made from a lipid (fatty) layer: you can imagine it like the outside of a washing-up-liquid bubble. This acts as a boundary between the substances inside the cell (proteins, chemicals, the cell's own DNA) and the nutritious broth, or growth medium, that the cells are grown in. The DNA for the vaccine cannot get across that membrane into the cells. So we put it in a solution that coats it in a bubble, similar to the lipid layer that makes up the outside of the HEK293 cell. When the vaccine DNA in its solution is then mixed with the HEK293 cells, the two lipid surfaces can fuse together, in the same way as when two small washing-up-liquid bubbles fuse together to become one big bubble. The DNA is now inside the cell, where it provides the instructions for the cell to make lots of copies of the adenovirus.

We always make lots of these vaccine DNA-and-HEK293-cell preparations in parallel, because some of our synthetic DNA will contain errors. This time around, because we had never used this method before, we also made preparations with lots of different conditions: different ratios of adenovirus DNA to spike protein DNA, different ratios of cells to DNA, and so on.

After about a week we should have started to see that virus particles were being produced inside the HEK293 cells. (Each

* We will say more about these elsewhere but the important thing to know about them here is that they contain the adenovirus gene E1. This gene allows the ChAdOx1 adenovirus – which has had its E1 gene removed so that it cannot replicate itself inside normal human cells and cause an infection when used as a vaccine – to replicate itself inside these cells.

individual copy of a virus is called a virus particle.) But by 17 February, nearly three weeks later, we could not detect virus particles in any of the dozens of preparations we had made. The rapid method had failed.

This was obviously a setback and a disappointment – we were aware that Sarah and other colleagues were waiting for us. But we didn't take it too hard. We had not had the time to do any practice runs and it was always a big ask to expect a new method to work first time. (The plan is to go back to this method and figure out what went wrong so that we are ready for next time.) Fortunately, while we had been trying to get the rapid method working, Sue had also been working on a backup, using the slower but much-practised classic method. So on 20 February, as families were enjoying half-term ski holidays in Austria and Italy, unaware that they would be bringing coronavirus back home with their memories of the mountains, we activated plan B: the classic method, but faster. Classic plus, you might call it.

Traditionally we do the transfection then leave the mixture incubating for up to a week at this stage, to allow viruses to be made and to proliferate. We then purify the vaccine out of the cells it has grown in. This means, we remove the complicated mixture of proteins, nucleic acids, fats and carbohydrates that make up the contents of a human cell so that we are left with pure viral vaccine particles. After that, we clone the viral vaccine particles, meaning we isolate single virus particles, and allow each one to infect a tiny cell culture and make lots more copies of itself inside those cells. ('Culture' is our term for cells in growth medium.) Then we check the genetic sequence of each clone and proceed with whichever ones are completely correct and homogeneous. This lengthy and labour-intensive cloning process is known as single-virion cloning (because we are making a clone from a single virus particle, or virion). It is necessary to ensure we end up with a vaccine that contains completely genetically correct and identical viral particles.

There are always some errors in the strands of synthetic DNA provided. The DNA is made by a machine using chemical reactions, and no chemical reaction can be 100% accurate when you are repeating it 100 billion times. Also, mutations or contamination can happen during the manipulations we do to get this far. Over the years we have found many different ways that things can go wrong during this whole process that are only of interest to the nerdiest of virologists. Single-virion cloning, though it was slow, labour-intensive, expensive, and still did not always work, had always been the solution.

But this time, to go faster with a better chance of a successful outcome, we came up with an alternative plan: the 'classic plus' plan. In any GMP manufacturing facility, there is what is called a 'qualified person' or QP – this is the person who ultimately certifies the release of every batch of product. It is obviously a role with great personal responsibility. Eleanor Berrie had been a QP for more than twenty years and had enormous expertise in adenovirus production. Eleanor's plan was, instead of working from single virus particles, which involved incubating for a week, purifying, isolating virus particles and cloning them, we would work from single infected cells. After just one day of incubation, we would take the population of cells that had been transfected, dilute them in order to separate them into individual cells, and introduce each of those individual cells into a new, uninfected culture. Each single infected cell would then produce an independent preparation of the vaccine, which could be expanded and tested to check it was completely correct and homogeneous – that is, a true clone.

Limiting dilution – working from individual cells – was a much faster but less reliable way of getting a completely correct and homogeneous population than single-virion cloning – working from individual virus particles. Our timing here was fortunate – it was only possible to use this method because of recent advances in DNA sequencing. At the CBF, we had been working for about a year to implement a technique called next-generation sequencing (NGS), which allowed us to read the whole genome sequence of any adenovirus ourselves, in twenty-four hours, at high sequencing depth, for around £500. We could therefore be confident in proceeding with material that had not been through the traditional, more reliable but much slower procedure, because we could check multiple preparations and any erroneous or mixed preparations could be reliably detected and discarded. This would save us weeks if not months of effort – and, more importantly here, of time.

Eleanor set up the 'classic plus' transfection on 20 February, three days after we decided the rapid method had failed. She would not normally be doing lab work: this was like getting the professor to mop the floors. But while the rest of the country was maybe not taking this new virus seriously – our prime minister had yet to attend a COBRA meeting on the topic – we were following the situation in China and in Europe. We knew the stakes were high and we could not afford another failure. We needed Eleanor's hands-on expertise.

By the next week we could see that some of the preparations did contain viral particles. You can't see these with the naked eye, or even with a standard microscope. But the cells – which are much bigger than the viral particles – grow in a single layer on the surface of the plate. What we could see down the microscope was holes in the cell layer, where a cell had become infected, burst, and gone on to infect its neighbours. We had clones!

From that point we needed to expand these tiny amounts of viral vaccine, to generate enough material to sequence their

DNA. We do this gradually, moving our viral vaccine to ever-larger containers: from an indentation the size of the rubber on the end of a pencil, to one the size of a two-pound coin, and so on, until a few days later our five most promising-looking candidates were each transferred to a flask with a surface area of 175 cm^2.

At that point we sequenced the candidates using NGS and discarded any that were not 100% genetically correct. By early March we had picked a favourite that was genetically perfect and proliferating particularly well: its name was D8 and I have a lovely picture of it at four days old. It looked like a tiny heart.

We needed enough starting material to seed the manufacturing process many times over, so we now set about making a much larger preparation of D8. To do this, we used special flasks called hyperflasks, each with a surface area of 1,750 cm^2, equivalent to about three pages of A4 paper (the cells grow in a single layer on the surface area of their container). We seeded one hyperflask of cell culture with some of the D8 vaccine preparation. The adenovirus vaccine infected the cells, turning them into adenovirus vaccine microfactories and killing the cells in the process. The next day we repeated the process, now using the amplified D8 adenovirus vaccine from the first hyperflask to seed eleven more hyperflasks of cell culture. From these we generated nearly 200 ml of starting material containing about 2 billion infectious virus particles per millilitre.

In total we ended up making about 300 ml of this precious fluid – just over half a pint, or a big mug of tea (we later repeated the final step to make a bit more). These few millilitres were destined to seed the manufacture of every dose of the Oxford vaccine ever produced. It seems incredible to me that the current plan is that there will be at least 3 billion doses, all born from this same starting material.

Step 2: Use the starting material to produce the required quantity of vaccine

Originally, the plan was for us at the CBF only to go as far as Step 1. As we were getting to work on the starting material, Sarah was in touch with the Italian company Advent, to arrange manufacturing there. The idea was that they would use our starting material to make the vaccine for clinical trials, leaving us free to continue with our other work (making a new vaccine against Ebola for Tess).

In the few weeks since that January meeting in Sarah's office, though, the world had changed quickly. By the middle of February the virus had seeded itself in dozens of countries all over the world. Although the WHO was still saying that it was impossible to predict what direction the epidemic (not yet officially a pandemic!) would take, it had also started discussing the possibility of making a vaccine within eighteen months. The critical piece of new information was that people with the virus could be asymptomatic and silently transmitting for days. The world was going to need this vaccine as soon as humanly possible. There was growing pressure for us to drop Ebola and immediately switch to making the Covid vaccine: the reality was that this would be the fastest way to get clinical-grade vaccine for first-in-human trials. But there was no money for it, and we hadn't finished the work on the Ebola vaccine.

Feelings were running high. A group of key people came together for a crucial meeting in early March: me, Sarah, Tess, Andy Pollard (designing and eventually running the trials), Sandy Douglas (already thinking about scaled-up manufacturing) and Adrian Hill (head of the Jenner Institute and already encouraging us to consider commercial partnerships and eventual distribution to low- and middle-income countries). We decided that as soon as the starting material was ready, my team would drop everything else and start production, here in Oxford.

I have to admit that coming out of that meeting I felt both exhilarated and terrified as I braced myself to tell my team that there had been a change of plan, and that all eyes would be on us as we attempted to make a vaccine in record time, with no backup plan, and no room for error. I reassured myself: they are experts, they know what they are doing, I have faith in them. The Covid-19 vaccine project could not be in better hands.

The CBF's most senior expertise in adenovirus production lies in three dedicated women: Eleanor Berrie, our senior QP; Emma Bolam, our head of production; and Cathy Oliveira, our new production manager. The four of us immediately got together to work up a plan. How much did we realistically think we could make, and how fast? We made three decisions.

Our first decision was that we would continue to proceed 'at risk', moving to the next stage of the process before all testing was complete on the previous stage. Again, to be clear the risk here was not to the end product, but to us. For example, we were expecting the starting material to be ready by 17 March. We would never in normal circumstances consider using a starting material until it had undergone a whole suite of tests to be sure that it met strict standards of purity and quality. But these were not normal circumstances. To move quickly, we would still perform all the same tests as usual, we just wouldn't wait for the results before moving on to the next part of the process. If the starting material failed any of its tests, we would have to throw out anything we had made from it. But that risk – a risk of wasted time and effort and serious money, but not of quality – was one we were prepared to take.

Second decision: the Ebola vaccine. We just needed to get

it to a stage from which we would be able to restart the work later. The production team, headed by Cathy, went rapidly to work, purifying the Ebola vaccine material and depositing it safely in the deep freeze.

Thirdly, we had to make a decision about preparing our clean room. The measures we take to reduce the chance of contamination during manufacturing include very efficient filtering of the air inside the room, frequent air changes, and protective clothing (overalls, overshoes, gloves, hats and goggles) for anyone working inside the room. At the CBF we usually also strip-clean and fumigate the entire suite between manufacturing projects. This is not strictly necessary, but it is our standard practice.

The suite *had* been fully cleaned and fumigated just before we started the Ebola vaccine manufacture in January. The four of us thought very carefully about the risks that would come with skipping the fumigation in this case. The biggest risk (albeit a very small one) was that the Covid-19 vaccine might get contaminated with a bit of the previously manufactured product.* We had a very sensitive and specific test for this valuable product, so we knew we would be able to test our final vaccine to check if there had been any contamination. Proceeding without fumigation would save at least three weeks. We drew up a formal risk assessment and submitted it to the Medicines and Healthcare products Regulatory Agency (MHRA, the body responsible for approving every step of our vaccine development process and ultimately for deciding whether to allow it to be used), who agreed our approach. (This was the first of a very large number of communications

* Just a reminder: that product was an adenovirus vector with a bit of the Ebola gene sequence in it. The vaccine did not contain Ebola virus and we never handled Ebola virus.

we would have with the MHRA over the coming months. That relationship, and the MHRA's proactive approach, is a critical part of this story.)

To make the vaccine, you need to introduce the starting material into the human cells that will act as the virus-production factories. When we were making the original starting material, we grew the production cells attached to plastic plates or flasks in a neat layer, like a tiled floor or patio. That was useful because it meant later, we could monitor whether the virus-production process was working by looking down a microscope and checking for a hole in the cell layer. Cells that were infected by the virus would make so many copies of the virus that eventually they would burst and infect adjacent cells, creating a visible hole. But this isn't a practical approach for large-scale manufacturing. You would just need far too much surface area. Instead we use cells that have been adapted so that they can grow suspended in their nourishing broth in constantly moving conical flasks. At the CBF we had a small supply of these specially adapted HEK293 cells, which we had grown ourselves, tested very stringently, then preserved in small vials in a very low-temperature freezer. But to make lots of vaccine we needed lots of these production cells, so we needed to expand them.

On 6 March, our new production manager Cathy took two of our precious vials of HEK293 cells out of the frozen store. Once the cells were thawed she introduced them into a small conical flask of growth medium. Over the next two weeks, the cells grew exponentially: that is, in the first day or so they doubled, then in the next few days they doubled again, and so on. As the cells grew, we kept increasing the amount of broth we were growing them in, and moving them to bigger and more containers. At the start we had two tiny vials' worth of cells. Two weeks later we had nearly ten litres of cell culture, ready to receive the virus starting material.

At the same time as our production cells were growing exponentially in the lab, so were case numbers out in the world. And the consequences for our daily lives were doubling and redoubling too, with things getting exponentially weirder and worse every day. This is how exponential growth works. The first doubling, from tiny to small, looks like nothing. After a few more doublings, though, it's amazing where you get to. On 10 March, as Sarah headed off to the House of Commons, the whole CBF team were in a meeting on infection control: hand washing, and something new to us called social distancing. There had now been nearly 300 confirmed cases in the UK. I needed the members of my small team not to get the virus, and definitely not to pass it on to their colleagues, because there was nobody else who could do this work. There was no way we could deliver this impossible task to its impossible deadline if my team started to get sick.

Two days later, the government issued advice that anyone with a continuous cough or a fever should self-isolate for seven days. People aged over 70 or with pre-existing medical conditions were advised to avoid cruises (February had been full of headlines about the *Diamond Princess*, the cruise ship with 3,700 people on board that had to quarantine passengers in their cabins and at one point accounted for half of the world's confirmed coronavirus cases outside China). Another two days after that, UK retailers issued a joint letter asking customers not to panic-buy as some shops were running out of toilet paper, pasta and hand sanitiser. By 16 March all my friends were convinced they had coronavirus but no one could get a test. The shop shelves were emptying fast and I couldn't get a delivery slot with Tesco.

On 17 March the Foreign Office advised against all non-essential travel and it was impossible to get hold of hand sanitiser anywhere. I made our first own-brand batch, using lab chemicals according to the WHO recipe. It was thin and smelt of alcohol but it would work. We repurposed a soap dispenser and installed it by the front door of the lab.

On 18 March the government confirmed that schools would close indefinitely from that Friday, which was in two days' time and many of us – including me – panicked about what we were going to do. I quickly secured letters for all my team members with children confirming that they were key workers and so their children could continue to go to school. There were no eggs in the shops and it was announced that as of that evening pubs, cafés and restaurants would have to close. There was lots of chat amongst my friends about meeting up while we still could, and I had to give them a talking to.

Sunday 22 March was a pretty strange day. Two team members emailed to say they felt unwell and would not come to work the next day just in case. On the upside, my daughter Ellie brought me a cup of coffee in bed (it was Mother's Day), I was invited to contribute to the government's Vaccines Taskforce, and my favourite pub the Magdalen Arms was doing takeaway for Sunday lunch.

Two weeks after Cathy had introduced our thawed HEK293 cells into a small conical flask, we had about ten litres of culture, containing 20 billion cells. On 23 March we inoculated our ten litres of culture using about ten millilitres of our starting material. This was a big moment. We were putting the bread in the oven, and then we would just have to wait to see whether it was going to rise.

I think it was that day that I first felt the fear: a low twist at the base of my guts that said, 'this is madness, there is no way this is going to work'. Blimey, that was a crazy day. My phone was buzzing with updates and queries non-stop, and although I was interviewing DPhil candidates all day, in every moment between interviews I pulled out my laptop and rattled off a few responses. Keeping the non-vaccine part of my job going in 2020 was really difficult and would have been impossible without supportive and understanding colleagues.* That evening, my guts twisted again as the prime minister announced a nationwide lockdown and told us we must stay at home. I had to give myself a talking to: no matter how much I was freaking out, the team couldn't see it. They needed me calm.

Two days later, our bread was ready. Our production cells had produced lots of virus but not yet burst. The harvesting could begin.

Step 3: Purify the vaccine

In this step, we get rid of all of the bits of the human cells in which we have grown our viral vaccine, and any badly formed non-functional viral particles, leaving us with pure vaccine. We want the bread, not the tin we baked it in. Or, in other food terms, it's like having a vast vat of minestrone – only on a microscopic scale – but only wanting the carrots. This is by far the most difficult part of the process.

We again had a choice to make about the method we would use. This time it was between using a well-rehearsed method that we were confident would work but that would produce less vaccine (about 500 doses), or a new method that might go wrong but if it went right would yield much more product (maybe 1,000–1,500 doses). By now the clinical-trials team was

* To those people, I owe you some fizzy wine when we are through this.

starting to think big and Andy was asking me pretty much every day how many doses we would be able to make. I always had the sneaking suspicion that whatever number I gave him, he added 10% in his head, without even realising it. Or maybe on purpose, just to push us harder.

The CBF manufactured its first adenovirus-vectored vaccine in 2007 (against malaria – a vaccine that is *still* undergoing clinical trials). Ever since then, we had used the traditional purification method. At the time of harvest, most of the adenovirus vaccine particles are still inside the HEK293 cells. The cells are relatively large so they can easily be separated from their nutrient broth by a low-speed centrifugation. The culture of vaccine-containing cells is simply transferred to sterile bottles, which are spun at 3,000 rpm in a centrifuge – effectively just a large, high-powered salad spinner. All the cells containing the vaccine form a pellet at the bottom of each bottle, and the broth can simply be poured away. In fact you could just leave the flasks for an hour and the cells would all settle to the bottom: centrifugation simply speeds things up.

We need the vaccine to be pure. We don't want traces of the production cells in it, not least because those vaccinated would then make immune responses against components of the HEK293 cells instead of against the spike protein. To get the vaccine out from the pelleted cells, we add a lysis buffer (a solution of salts and detergents) and 'pop' the cells by freezing and thawing them three times. As the cells freeze, the liquid inside them forms into ice and expands, bursting the cells' external membrane. The contents – including the virus particles – leak out into the buffer solution. We then thaw and recentrifuge and because the vaccine particles are very small they stay in the solution while the broken bits of cell form a new pellet at the bottom of the tube. Now we have a solution called clarified cell lysate – this contains all the properly formed virus particles, but also any empty adenovirus shells that have not formed properly and will be useless as vaccines, as well as other soluble material that came out from the cells when they burst.

To purify the active vaccine, we use the fact that the particles we want have a different density from all the bits we don't want. We make two salty solutions of different densities and carefully layer them inside a specially strengthened test tube. Then we carefully layer the clarified cell lysate containing our vaccine particles on top. It's a bit like the way a bartender makes a fancy layered cocktail, except that all the layers are colourless solutions and there is more pressure not to mess it up. We don't throw the test tubes in the air and catch them behind our backs.

Instead, we centrifuge the cocktail at a very high speed. The G-force that the mixture is subjected to is so strong that the different tiny particles in the solution separate into the layers, settling where they are just about floating. Denser particles need a saltier solution to float (think about the salty Dead Sea), so they settle in the lower, denser, saltier layer. The highest G-force you can experience on a roller coaster is currently 6.3. Most humans would be unconscious at 9 G. When we spin our test tubes to collect our cells, they reach 200 G, and this final spin to purify the vaccine is at 154,000 G.

After two hours of spinning, all the properly formed vaccine particles are gathered together in a single band within the tube, with the empty, unwanted particles and other bits we don't want in another layer higher up the tube. We collect the liquid containing our vaccine particles using a needle and syringe inserted through the side of the tube at the point where the vaccine-containing band has formed, and then repeat the centrifugation process to make sure no empty particles have snuck through. Then there is one final purification step to remove any remaining salt.*

The traditional purification method I've just described is very effective and my team is very experienced in doing it. But it is labour-intensive, takes a long time, and is not easily automated or scalable. We also end up discarding a significant amount of good virus particles during the process. None of that matters normally, when we are required to make a small high-quality batch of a new product for clinical trials, but now, with the planned trial clearly promising to be one of the largest we had ever run, we were going to need all the vaccine we could get, and it did matter.

A couple of years earlier we had made a rabies vaccine for our colleague Sandy Douglas. This was a project designed not only to manufacture a new effective rabies vaccine but also to develop improvements to the adenovirus vaccine manufacturing process for all vaccines using the ChAdOx platform. The work done on this project, successfully implementing a new purification process, laid the foundation for all of the large-scale manufacturing that was to come.† In contrast with the traditional method, the new method was fast, automatable, scalable and reproducible. It also yielded double or triple the amount of vaccine. But we had only done it once.

Going with what we knew was the safest option, but could result in a lot less vaccine being produced. On the other hand, if the new method went wrong and we had to start all over again making a new batch of vaccine it would set us back weeks.

* To do this we use a process of dialysis that works on the same principle as the dialysis given to kidney patients to remove the waste products from their blood.

† We had also looked at how to make the vaccine stable at room temperature, which is something we did not succeed in then, but will come back to.

It was a hard decision and once again discussions got heated. My team are generally cautious, which is exactly what you want in this job: a careful crew who will do everything possible to make sure our operations are compliant with the rules. I had to listen to their concerns about the risks. But we were also all very aware that maximising the doses so that the trials could be as big as possible was going to be important. As we did so many times throughout making the Covid vaccine, we took the more unknown path that was higher risk, but higher reward, alongside the tried-and-tested method. Obviously this was double the work for the team, but we had learnt by now that we had to hedge our bets and always have a backup. We decided to split the batch – two of the seven flasks would be purified with the new method, and five with the traditional one. The subsequent meeting with Sandy was for once shorter than planned – no rows, no one threatened to punch anyone on the nose, everyone was happy.

Purification is always a nervous few days in the clean room. The vaccine is all contained in a few flasks, then a few tubes, and finally a single bottle or bag, and a mistake at any point can jeopardise everything. You need a steady hand and a calm temperament. Fortunately the team had both.

On 27 March Cathy opened the centrifuge, removed the tubes that had gone through the traditional purification method and took a photo for our records. The photo showed an enormous fuzzy cloud of vaccine right in the centre of the tube. Later that day she emailed it to the whole team, with the subject line 'Look at these babies!' She said it looked like a halo. Our yield looked good and the product looked perfect. Someone emailed back, only half-joking: 'This might just be the fuzzy band that saved the world.' It was a big relief to have got the first batch safely to this stage – to have successfully turned the bread out of the tin – and the whole team was on a high. I think this was the moment we believed we might pull this off.

The contents of all the tubes were pooled, the final purification step was performed to get rid of any remaining salt, and our first batch of vaccine – in total a bit less than 500 ml, less than a pint of milk – was put into a sterile bag and into the freezer, ready for the next step: putting the vaccine into vials.

Then we went back to process the second batch using the new method. At first everything seemed to go well. We had apparently good yields and the first step of the purification went as expected. But on the evening of 8 April I opened a devastating email from Cathy. The material that was coming off the final step was behaving very strangely. Instead of the expected perfectly clear solution, the liquid emerging was cloudy. We knew this was not a good sign. On the phone to Cathy a few minutes later, the atmosphere at the lab was subdued. We had never seen this before and no one really knew what to do. It was late at night so we agreed that we should put everything safely in the fridge, everyone should go home and try to get some sleep, and we would approach it with fresh brains in the morning.

The next morning we collected all the material and ran some tests. Very little vaccine could be detected in the final product. We spent the rest of the day trying to work out whether there was some way to salvage something from what we had, but by the evening it was clear that it was hopeless. Our best guess was that the virus particles had all clumped together somewhere unexpected at the final stage and been accidentally discarded.

I don't think I can explain how awful this felt for the team. They knew how important this was and they felt personally responsible for the failure. That night I gave everyone a pep talk: they had already done one impossible thing, I told them, creating the first batch in record time. Asking for two impossible things was just too much. Even to have tried was enough. I reminded them, as well, that it was because of their reasoned arguments at the start that we had split the batch in two and still had all the excellent material from the first purification run. Imagine the disaster if we had just gone all-in for the new method and the one shot had failed. Yes, this run had not been successful, and yes we had lost some material. But we hadn't lost all of it. They should remain proud of their enormous achievement, and not get dejected. We had to pick ourselves up and keep on going.

Step 4: Fill the vials

Because we are going to put it into people's arms, the final vaccine product of course must be completely sterile. Clean-room areas are divided into grades. The highest grade, A, is a completely sterile environment, so there can be no bacteria or other microorganisms present in the air or on the surfaces. Obviously this is difficult to achieve. Microorganisms are all around us, in the air, in the soil and on our bodies. But this sterile environment is what we need for filling the vaccine into vials.

Because we only make small batches, the simplest and most cost-effective way for us to fill our product into vials is manually, inside a grade A isolator which is itself inside a grade C clean room. The isolator is a sealed box, the width of the room, into which completely purified air is pumped to keep it sterile. One side is a glass screen with two pairs of sealed gloves set into it so that the operators, sitting side by side outside the box in their full protective gear, can handle the materials and equipment inside the box without touching it with their hands. There is a hatch at each end so that wrapped, sterilised materials and equipment can be placed inside the box from one end, and filled sterile vials can be removed from the other. Vial filling is one of the most skilled jobs that we do at the CBF. It is high-pressure work that is also boring, repetitive, delicate and tiring on the arms. The production team know how to handle it though and are very steady.

On 28 March we had deposited our first successfully purified batch in the freezer. A couple of days later I did a radio interview in which our work was introduced as 'the only story in the world'. My mind flashed to the lab. There was so much to do, and so much that could still go wrong. Several members of the team were having to self-isolate as they lived in shared houses with a housemate who had symptoms. Others were having to shield. Lots of us had kids on their Easter holidays, although the schools were staying open for the children of key workers, which was amazingly helpful. Being the only story in the world felt like quite a heavy weight to bear.

Thursday 2 April was fill day. Fill day is always a big day at the CBF. The whole team is involved as there is lots of checking, testing, monitoring, distribution and documentation to do. We removed our small bag of vaccine from the freezer, transferred it into the grade A isolator and then pumped it through a series of filters with tiny pores that remove any bacteria. The final sterile filtered material was then ready for filling.

It wasn't tense exactly, but there was a definite feeling of focus and close attention. On other days it's often quite jolly, but today there was no chat. Inside the hush of the clean room, our filling operator for the day, Ioana Baleanu, sat head-down at the glass screen and, using a microdispenser system (a tiny, very precise syringe), carefully placed approximately 0.5 ml (a tenth of a teaspoon) of vaccine into each small sterile glass vial.

She then passed the vial, still inside the isolator, to Emma, the day's crimping operator, who delicately added a sterile rubber stopper and a metal crimp to keep it closed and put it on a rack. When the rack was full, it was passed out of the isolator through the hatch. Each vial was then inspected for defects, and vials that passed were logged and labelled. All the time, samples were being taken and the isolator environment being monitored for sterility.

Normally on a fill day we order in a big lunch for everyone, as there is no easy way for people to go out for food during a very busy and long day. This time was a bit different. There were no platters of sandwiches. Everything was individually wrapped and we ate socially distanced at our desks. This just brought home to us the strangeness of the situation and the hope that was tied up in what we were doing. By the end of the day the team had filled, logged and labelled 500 vials.

For the first time in ages, we felt like we could breathe more easily. Even though we realised we were only at base camp, it felt like we had a climbed a mountain. (And I hear it's quite difficult even to get to base camp on Mount Everest.)

Step 5: Label the vials, document the process, and test and certify the quality of the product

This is steady, behind-the-scenes work, checking everyone else's work and ensuring that strict standards are met.

Labelling might sound like a simple job, but there are very comprehensive regulations involving a lot of documentation to ensure that products for clinical trials are properly labelled. Helena Parracho (who is in charge of labelling) got volunteers from across the CBF teams to help her in getting everything assembled, checked and ready. A mistake at the labelling stage could result in a volunteer being given a wrong drug or a wrong dose, which of course could be very serious.

The documentation of the process is necessary in order to get approval from the MHRA, without which no trial can go ahead in the UK. We need to submit a dossier of documents describing the manufacturing process and the full trial plans. Every stage of the manufacturing process is governed by regulations, which we make sure we adhere to by having detailed protocols for everything. The process of putting on protective clothing before entering the clean room, for example, is described in a nineteen-page document, with another five documents totalling sixty pages relating to the supply of the bunny suits, hair coverings, etc. Each person working in the clean room has to be trained in the procedure, and the training is repeated every six months. There are some people who unfortunately through no fault of their own are heavy shedders of the bacteria we all carry around with us, and those people can't work in a clean room at all. As another example, putting vaccine into vials can only be done by people who have been trained, and have then performed the procedure on the number of vials we expect to fill, using sterile broth instead of vaccine, on three separate occasions. The vials of broth are then put in an incubator to see if any bacteria grow. If they do, that means the operator has failed to maintain sterility, and they have to do more training and repeat the test. Once they pass the test, they have to retake it every six months. We have detailed procedures to make sure everything has been properly planned (quality assurance) and then lots more to test that everything went as it should have done (quality control). And records are kept for all of this. If it isn't written down, it didn't happen.

Once the product is in vials it also has to undergo a series of stringent tests to ensure that it is completely sterile and uncontaminated and of appropriate quality for use. After all three of those things have happened, we can take the final step of certifying that the batch produced is suitable for use in a specific trial.

Testing is not glamorous or exciting, but it is obviously a crucial part of the process. It is also a very important part of this story because of how much time we managed to save. From very early in the year we all knew that every day we could shave off the process would save lives. We were constantly looking to see how we could do things one day faster, two days faster, without compromising safety. And in the end, we managed to shave more than two months off our testing time.

We have to send some vials off to specialist companies for parts of the testing, and these external tests had in the past taken three months or more. This time, before we had even started manufacturing we had been in intense discussion with the MHRA about how we could speed up testing without compromising on safety. We had devised a rapid-testing programme, which was much more expensive, but also much faster.

As soon as the vaccine was available we initiated the rapid testing. Proceeding 'at risk', we also immediately started preparing all of the documentation the MHRA would need, and labelling the vials. Normally we would wait to do this, as there is no point wasting time preparing documents and labelling the material if the tests fail: we would just have to delete the files and throw away labelled vials. But in this case we needed to be ready to use the vials the moment we had the all-clear on the tests and the approvals.

With a lot of late nights and weekend working, and masses of support from our testing partners and our specialist couriers (we shipped the vaccine frozen on dry ice and our biggest fear was that there would be a courier hold-up and the dry ice would run out and the vaccine would thaw, invalidating the tests), we managed to reduce our test time from three months to just over two weeks.

We got the final test results back on 21 April, and the next day our QP Richard Tarrant certified the batch as ready for use. The whole team watched while he signed the final document that would mean the product was ready to go. The vaccine was 'shipped' – in reality just carried across the road – to the clinic, ready for the first vaccination to take place the next day.

We had gone from DNA construct to clinical trial in sixty-five days – an achievement no one would have thought possible in February when we started. We were all exhausted. We had a slice of socially distanced cake and went home to try to get some sleep.

At this point no one knew for sure whether the vaccine would work, but even so I think we all felt that we had done something enormous.

CHAPTER 6

Scale-up

1 February–30 June 2020
Confirmed cases: 12,038–10.46 million
Confirmed deaths: 259–508,179

Thursday 30 April. It was a strange time. The elderflowers were out. I know that because I was swapping tips with my friend Sally about the best biomanufacturing process to make them into home-made fizz. And the days were getting longer. I know that because it was light in my kitchen when Andy's email came through at 7.10 a.m. Spring was coming but it felt like we had been in lockdown forever, and case numbers were still going up, and everything was still awful and there was no end in sight. I read Andy's mail:

> The University of Oxford has today announced an agreement with the UK-based global biopharmaceutical company AstraZeneca for the further development, large-scale manufacture and extensive distribution of the Covid-19 vaccine currently being trialled by the university. This is an important development because it will provide the critical infrastructure needed to ensure global equitable access to the vaccine, if the clinical

trial demonstrates that the vaccine is safe and effective. The trial has just begun and is progressing well.

Over the previous month it had become clear to all of us working on it that the project was rapidly getting to the point that we at the university couldn't handle it on our own. Since January, my team had been working flat out to lovingly hand-make the first few hundred vials of vaccine for the trials. But while it had been a big achievement to get these first vials ready in record time, if the vaccine worked we were going to need a lot more. A lot, lot more.

How do you go from a few hundred vials, to millions, perhaps even billions? It's a huge scale-up in a short space of time. My colleague Sandy Douglas and his team had been developing a process that we were confident would be capable of producing millions of doses. But the logistics of actually making them, and then distributing them across the world, were daunting.

I had known that negotiations were taking place with at least one major pharmaceutical company. I also knew that there had been discussions about whether the Vaccines Manufacturing and Innovation Centre – an organisation that was still being established, but that once fully functioning will be a centre of excellence for vaccine development and manufacturing, and provide the UK with a national vaccine response capability – could take on the role. (They were not prepared to do so – wisely, it turns out. I don't think any of us realised what an enormous undertaking this would be and with hindsight we could never have achieved it without big pharma.) But Andy's email was the first time I heard mention that it would be AstraZeneca, and it was something of a surprise. I knew they were big in cancer medicines and they were obviously a name in the pharma world, but they did not have a particular reputation for vaccine manufacture.

We felt a bit disconnected: as though decisions that would really affect our working lives (by this point all of us were working on this project all of the time and it had completely taken over every waking and sleeping hour) were being taken at the highest level of the university with no consultation with those of us who actually knew how to make this vaccine. We also felt a bit of trepidation that teaming up with a pharmaceutical giant would mean someone else making large profits off the back of our efforts. We were realistic that there would need to be a profit motive for big pharma to get involved, but we were hopeful that if the vaccine was going to end up making a lot of money for other people, at least the university might also see some returns on its investment. The press release that Andy had sent round suggested that both partners had agreed to operate on a not-for-profit basis for the duration of the pandemic, and that royalties returned to the university after that would be reinvested into pandemic preparedness. This was reassuring, as was the intention to commit to delivering the vaccine at low cost to the parts of the world that would not be able to pay for an expensive vaccine, which was something that all of us on the team felt strongly about. But it did feel like our relationship with the project had slightly shifted.

We were all, at this point, very tired. April had been a tough month, dealing with the start of the trial, making our first batch of vaccine at the CBF, liaising with Advent in Italy (whom we had contracted to make a second clinical trial batch for us), talking to the regulator, and working on scale-up. Like our colleagues in the NHS and in care homes, we were having trouble getting hold of PPE, and by early May I had so many staff off sick or isolating that I was working in the lab as well as everything else. We had been working flat out and were continuing to do so, and at first, having AstraZeneca on board felt like an additional problem rather than a solution. Looking back I realise how negative and unreasonable this sounds. We could not have got anywhere without them, and we all get along very well now. But at the time, when we were already working eighteen-hour days, we were not always happy about our demanding new team member. Especially when it felt like our new team member was actually our new boss. (For a while I changed my WhatsApp bio to 'working for the man'.)

AstraZeneca wanted to be hands-on immediately and the requests for meetings, for documentation and for samples started coming thick and fast. Early on, the meetings – all done remotely on our computers of course – felt slightly haphazard. AstraZeneca is an enormous entity, with multiple teams across the UK and the US with quite specialised roles, whereas everyone at our end was involved in and knew about everything. Also, they had no experience of manufacturing viral vectors, so the technical aspects of producing viral vectors, and the quality tests needed for these kinds of products, were all new to them. It was frustrating to have to keep repeating ourselves to slightly different combinations of AstraZeneca people. Equally, they were so used to huge global projects that they didn't always realise how much we didn't know about the logistics of managing large-scale manufacturing or trials or distribution.

We were essentially a family-run pizzeria, doing everything ourselves, from checking with the clinics how many doses they needed each day to sticking on labels and arranging couriers, and they were Pizza Express, with software and huge systems and outsourcing to run global operations. They also used a lot of abbreviations that we didn't understand, and we sometimes felt foolish having to ask them to explain what they were talking about. They must have thought we were so basic. So there was a steep learning curve on both sides of the deal in the first month.

But it was also clear from the start that AstraZeneca were thinking big – bigger than I had imagined at that point – and hearing the scale of their ambition was pretty inspiring to us all. I remember being in a meeting very early on, probably in May, when someone at AstraZeneca confidently used the phrase 'billions of doses'. That's a real shock to the system when a really big day for you is manually putting 500 doses into vials. They were prepared to throw everything at it straightaway, rather than waiting for results from clinical trials before they fully invested.

The AstraZeneca team also immediately pushed the project in what turned out to be a vital direction that would make a huge difference to who could receive the vaccine around the world. When we make small batches of ChAdOx1 products for clinical trials at the CBF, we always store the final product at -80 degrees centigrade. Ultra-low-temperature freezers are routine pieces of kit in research laboratories and clinical-trial sites, and storage in these conditions means that our limited quantities of precious trial material have a long shelf life (in some cases more than five years). We did however have a lot of data for many of our existing adenovirus vaccines that demonstrated that the material was stable at 2–8 degrees, normal fridge temperature, for at least a year. When AstraZeneca saw this data they quickly understood its importance. They immediately initiated studies on the first large-scale batches of ChAdOx1 nCoV-19 coming through. The results, showing that our vaccine could be transported and stored refrigerated rather than frozen, had enormous implications for how easy the vaccine would be to distribute and to use. Robust vaccination programmes for polio, measles and other childhood diseases, reaching every corner of the world, already needed refrigeration, all the way from the factory to a tiny healthcare facility to a person's arm. So our vaccine would fit into existing, tried-and-tested supply chains and would be able to travel far and wide. We wouldn't have to worry about the vaccine thawing during transit, or about whether there was a reliable freezer – let alone an ultra-low-temperature freezer – at its final destination.

But I am getting ahead of myself . . .

—

In the early spring, while others were making initial contact with AstraZeneca and drawing the road map to billions of doses, my day-to-day efforts were still at a rather smaller scale. We would need hundreds and then thousands of doses for our clinical trials, and the CBF-manufactured material that we were in the process of making (in the end we managed to get about 700 doses from the material we had purified and frozen at the end of March) would soon run out. While we were producing our own vaccine, and in constant discussion with each other and the regulators about how we could do that as quickly as possible, we were also working intensely with our Italian colleagues at Advent, and pulling together a group of manufacturers to help us to work out how to scale up way beyond what either we or Advent were capable of.

Right at the start, the original plan had been that we at the CBF would produce the starting material and then ship it to Advent for manufacture. However, in early March, when we realised that my team at the CBF would be able to complete production quicker than Advent, we changed the plan. We would make a first batch at the CBF, and Advent would stick to their plans and, using our starting material, make a second batch. We would definitely need it all: the trials were growing all the time, with more people and locations being added on an almost daily basis.

So half of the first ChAdOx1 nCoV-19 starting material went to my production team, for us to make the first batch of vaccine, and we sent the other half to Italy. That was on 17 March. We were still awaiting test results, but as was becoming normal we proceeded at risk.

Under normal circumstances, on receipt of starting material, an organisation like Advent would first do a practice run-through. But of course, this takes time. In February, as we were making our starting material, we had also been trying to persuade Advent to drop the practice run. It was uncomfortable at first – Advent had a reputation for quality that they did not want to risk. At the same time, they understood the urgency. Eventually they agreed to try to deliver to a very aggressive timescale, and we agreed to pay even if the run was unsuccessful.

Working with the team at Advent was one of the highlights of this project for me. They are a small team, about the same size as the CBF, and like us they specialise in viral vector manufacture. Their boss, Stefania Di Marco, is hugely knowledgeable. The situation in Italy at the time also meant that the whole team was particularly keen to help. The epidemic had taken hold there early. In late February, there had been a cluster of cases in Lombardy and ten towns had been locked down. Schools in the area had closed, supermarket shelves had emptied, Serie A football games had been cancelled. While UK deaths remained in single figures, Italy's soared past 1,000 and exhausted Italian doctors took to social media to try to explain what was happening there, and warn it would soon be happening elsewhere.[1] In early March, when there were no restrictions in the UK and no one was wearing PPE, the Advent team were only allowed to travel between home and work, and on video calls they were all wearing masks (and were shocked that we weren't). Their commitment was apparent, as was the challenge of trying to deliver against this stressful backdrop. Seeing the position of our Italian colleagues, and feeling their personal desire for the project to succeed, made it feel like we were all part of the same effort. We had just enough starting material (sourdough starter, from Step 1 of the five-step process described in Chapter 5) for one batch at the CBF and one batch at Advent. There would be nothing left over so we each had just the one shot.

The starting material arrived in Italy on 20 March. It was a nervous wait. Flights to Italy were already disrupted and as always we were worried about the dry ice evaporating. If the material defrosted it would be useless, and we would lose at least two weeks going back to make more. We were monitoring the temperature from a tracer included in the packaging and there was relief all round when it arrived still safely frozen. Like us, Advent were already preparing their production cells – the cells in which they were going to grow their vaccine. At the CBF we expanded two precious vials of HEK293 cells to a final volume of ten litres, from which we eventually made 700 doses. Advent, who operated on a much bigger scale, were expanding their cells in large bags, aiming for one hundred litres from which they would be able to make thousands of doses. They successfully infected their production cells with our starting material (mixing the ingredients and baking the bread, Step 2) on 13 April.

Step 3 is purification: removing the bread from the tin. Advent's method for separating the pure vaccine from the unwanted cell debris was similar to the method developed by Sandy (that we had just failed with at the CBF). The Advent team were very experienced in making high-quality products using this approach and their purification was successful. By early May they had filled 3,000 vials for us to use in the ever-expanding clinical trials (Step 4) – an extraordinary achievement from such a small team.

Step 5 was labelling, documenting and testing the product. Labelling the vials (a manual process) was a big task, so we decided to split the work between us. Advent shipped some of the vials to us overnight on dry ice and everyone slept with their fingers crossed as all was once more in the hands of the couriers. With major travel disruption across Europe and most flights completely grounded, I had visions of the vials going round and round, unclaimed and alone on a baggage carousel in an empty arrivals hall. (Obviously I knew that in reality they would be in the freight section at the back, but the airports must have been quiet and lonely places at that time.) With the vials safely delivered to the CBF, the team inspected each one individually, rejecting any whose seals were not properly crimped or any vials that were not perfectly intact, and proceeded to get the labels on. The Advent-produced material was needed straight away in the phase II/III trials that were about to start in the UK. But because it had been manufactured using a different cell line from the one we had used at the CBF, and using a different purification method, we had to demonstrate batch-to-batch consistency before we could use it.

Most of the tests that Advent had performed on their batch were the same as those we had done on the CBF material (testing for sterility, testing for the presence of residual cellular material from the production cells, checking for the absence of any known viruses, etc.) but there was one test that had been done differently. Advent assesses the amount of vaccine in its batch, or the concentration, using a method called qPCR, while we at the CBF use spectrophotometry. These tests use different techniques. qPCR is a molecular genetic test, measuring the number of copies of viral DNA per millilitre. Spectrophotometry estimates the amount of viral DNA based on how much ultra-violet light the material absorbs. (Each viral particle contains the same amount of DNA, so measuring the viral DNA tells us how many viral particles there are.) To check that the two measurement methods were equivalent we had previously sent Advent a vial of our vaccine, for them to test it using their method. They reported the same concentration as we had determined using our method, so we were initially confident that the tests were interchangeable. But, just to be sure, we would do the same thing the other way round. We would check the concentration of the Advent batch using spectrophotometry.

To our surprise, we did not get the same concentration reading as the Italians had measured. Instead, our tests suggested that this new batch was more concentrated than the Advent team had thought. The difference we found was about twofold. This was not a problem we had ever come across before: we would never usually need to use batches from two different manufacturers in one clinical trial, and we were only doing it this time because this trial was bigger and more urgent than any we had ever done before. We weren't particularly concerned, though: measuring virus concentration is difficult, and the margin of error on these tests can be up to 50%. We also knew from previous clinical trials that adenovirus-vectored vaccines are well tolerated and effective at a broad range of different doses so high precision is not necessary.

Just to check that everything was OK, we sent some vaccine over to Alex Spencer at the Jenner Institute, a senior immunologist who does a lot of the preclinical studies on the vaccines developed there. Alex used vaccine from both batches (CBF and Advent) to vaccinate two sets of mice, and then monitored their immune responses. Two weeks later, she provided data that both batches induced good immune responses, as hoped, and that there were no significant differences in the responses to the two batches.

This was all good news. Both vaccines were immunogenic. But we still didn't have a comparison of the vaccine concentrations, which would have required a much larger experiment using multiple doses for each vaccine. And we still had no explanation for the difference between the two test results for Advent's batch of vaccine. This was the subject of many meetings, as we had to decide whether to proceed with vaccinating the phase II/III trial volunteers, and if so, how.

The volunteers in the first, phase I, trial had been given a dose of 5×10^{10} (50 billion) virus particles, determined using our CBF spectrophotometry test.* We decided that we would use the same method to determine the dose of the Advent material for the volunteers in the phase II/III trial. By using this approach, we knew that the dose would be both well tolerated and immunogenic, whichever of the two results was accurate. It also erred on the side of caution: if the CBF method turned out to be wrong and the value determined by qPCR was in fact more accurate, then we would have given a lower dose than anticipated, which would be safe. If, on the other hand, we based our dose on the Advent value as determined by qPCR, and then that turned out to be wrong, we would have given our volunteers twice as much vaccine as we intended and we would not risk doing that.† We checked this approach

* This is a standard dose for adenovirus-vectored vaccines, and had been used in many trials previously with good results, though 2.5×10^{10} has also worked well.

† Although it would still not have been a completely unreasonable dose to give: for example, it would have been equivalent to the high dose that Johnson & Johnson tested with their Ad26-vectored Covid-19 vaccine in their clinical trials.

with the MHRA, and they agreed. The safety of trial participants is always the MHRA's primary concern, and ours.

Thus the initial dose given to the first volunteers in the phase II/III trial at the end of May was 5×10^{10} viral particles per dose according to CBF spectrophotometry. Or it was roughly half that much – 2.2×10^{10} viral particles – by Advent's qPCR measurement.

The recruitment of volunteers to clinical trials is often quite a challenge. Many of the trials we run in Oxford are to try to find vaccines to prevent diseases that are endemic in other countries, so there is no obvious direct benefit to the participants. It is a big commitment to travel to a clinic to be vaccinated with an experimental medicine for the good of a community on the other side of the world. However, Covid-19 was completely different. The disease had fundamentally changed life in the UK – schools were closed, people's livelihoods were at risk, people were dying, alone. Within hours of announcing that we were recruiting volunteers for trials, we had thousands of applications. I think that all of us involved have been humbled by the generosity and commitment of the trial participants, turning up to numerous clinics for blood sampling, and taking weekly swab tests for months and months. It reinforces my belief that people are in the main good and generous and altruistic. It's always worth remembering that the vaccine wouldn't have been possible without them. If we didn't have people who were ready and willing like this, we wouldn't have been able to gather crucial data on our vaccine, and we wouldn't be vaccinating people in the millions, as we are now.

But because recruitment to the trial was going so well, we were using up the vaccine at a rapid rate. The first filled vials from Advent had already been shipped out to the eighteen clinical sites around the UK. This had been a massive logistical enterprise, packing up and shipping the precise number of vials needed for each site – not helped by the fact that the sites kept asking for more, more, more. For weeks, lines of courier vans would queue up each day outside the CBF, and the team were working absolutely flat out to ensure that each vial was properly labelled, distributed and accounted for. I was occasionally getting very frustrated with the trial clinicians who seemed to find it very easy to demand that we provided fifteen vials to Hull tomorrow, without understanding that I had neither a magic carpet nor an ability to produce vaccine out of thin air. There were some stressful meetings and I might have sworn with frustration in some of them – not at people, but at the situation.

At one point early in June, a week or so into the phase II/III trial, it became clear that there was not enough vaccine in the country to meet the needs of the trial sites. Some sites were recruiting volunteers faster than expected, sometimes doses got wasted if volunteers did not turn up, and also the clinicians kept adding more elements to the trial – thinking of additional things to test – and the numbers just kept on rising. Advent had filled and labelled their second batch of 1,000 vials by this time, but there was a problem. The vaccine was still in Italy, the trials were in the UK – and there were no commercial flights operating between the two. We were stuck. What we did next is a perfect illustration of how out of the ordinary the situation was. My operations manager Omar El-Muhanna, who had a lot of experience working in extreme emergency situations, suggested that we charter a plane. It felt very rock star, even though it was the vaccine travelling first class and not any of us. It turns out chartering a private jet costs around £20,000, normally well beyond the budget of a small academic clinical trial. But by this time we had the might of a global pharma company behind us. All those meetings with our AstraZeneca colleagues were starting to come good: we got permission to proceed. The jet arrived in London the next day with no passengers, just a large box of dry ice and 500 precious vials for next-day distribution across the UK. The trial must go on.

—

After a week or so of vaccinations with the first Advent batch, the team monitoring the data coming from the sites was surprised to note that the volunteers receiving this new batch were having fewer and milder reactions (slight soreness where the needle went in, fever for a day) than we had seen in the phase I trial using the CBF batch. We had all been continuing to think about why the two tests for concentration had given different results. In Oxford, we decided to set up some new tests to try to identify differences between the two batches.

In my research lab in my other job, at the Wellcome Centre for Human Genetics, I regularly use two techniques to understand how proteins function in cells. I thought that we could use these techniques to analyse how well the two batches managed to persuade cells to make spike proteins (which is of course what the vaccine has to do in a recipient's body in order to trigger the immune response). I set up a system in which I could separately infect two different flasks of human cells with the two vaccine batches and then monitor how much spike protein was produced. When we aligned the quantities of the two batches, assuming the CBF test results were correct, the Advent batch produced less spike protein. When we corrected and aligned the vaccines using the Advent qPCR concentrations, the vaccines produced highly similar quantities of spike. This data, alongside the data from the trial participants' reactions to the vaccines, suggested that the CBF test method might not be the most appropriate way to measure the concentration of the Advent batch.

What we concluded was that our spectrophotometric concentration test was accurate when we used our purification method. Advent's qPCR concentration test was accurate when using either their purification method or our purification method. So, when we had sent our vaccine to Advent, the two tests had given the same results. The issue appeared to be that our spectrophotometric test was not accurate when combined with Advent's purification method. We had been keeping the MHRA up to date with all of our findings, and now reported back again. They said that the volunteers who had already received what we now believed was a half-dose could remain in the trial, and agreed a plan to continue, using qPCR to measure concentration from now on.

It was this first set of volunteers, who received a lower dose (because we initially used the spectrophotometry method), who were responsible for the unexpected, dramatic and confusing finding in November 2020 – the reason we ended up with the different efficacies at 62%, 70% and 90%, depending on which groups of volunteers were analysed. Our data seemed to show that an initial half-dose followed by a full dose offered better protection than two standard doses. However, the number of people who had received this initial half-dose was small, so our confidence in the accuracy of the finding was lower than our confidence in other results. We said at the time that we wanted to gather more data and perform further analysis to understand it. When we did, we discovered that the situation was even more complicated than it had seemed.

—

It was one thing getting hundreds and then thousands of vials ready for trials, but there was also an intense focus on gearing up to make millions of doses of vaccine for potential deployment, should it prove to be safe and effective.

Usually, we would not even start to think about scale-up or deployment until we had completed phase II trials, which would be at least three or four years into development. This time, though, we had started thinking about it back in February, before we had even produced a single vial of vaccine or designed the phase I trials. Right from the start of the outbreak in Wuhan, Sandy had feared that this disease was more concerning than initial reports suggested. From early February – when there had been just a handful of confirmed cases in the UK – he pushed hard for us to take it very seriously and get ourselves in the position to make large amounts of vaccine should it be needed. (He also said we needed to set up emergency childcare arrangements in case schools closed, which we all thought was completely mad, and did not do.) Sandy was the first person to suggest that it would be possible to get a vaccine from design to deployment in less than a year if the effort was made and funding was provided. At the time the big names were all saying it would take at least two years, but Sandy doesn't take no for an answer. And partly for that reason, he was proved completely right.

How, then, did we get to the point of being able to manufacture millions of doses of a completely new vaccine, using a technology that we had only ever used to make batches of a few thousand doses?

The great advantage of the type of purification process that Sandy had asked us to use for the rabies vaccine (that failed for us at the CBF in March, and worked well for Advent) is that, unlike the method that did work for us at the CBF in March, it is inherently scalable. Whereas our method, involving centrifuges and freeze-thaw cycles and making cocktails, was manual and labour-intensive, and would require exponentially more skilled people and centrifuges to scale up, with Sandy's method, you can just use ever-larger vessels to grow ever-more cells to produce ever-more vaccine, and ever-larger columns or filters to purify the material. And you end up with much more product. At the CBF we can grow up to ten litres of culture at a time, from which we can make a few hundred doses. At Advent, they can grow one hundred litres of culture at a time, from which they can make a few thousand doses. Commercial vaccine producers use special 1,000-litre tanks, or bioreactors, to make several million doses per run.

And it was not only the purification process that Sandy had optimised. Over the previous year he had also been working in his research group to optimise vaccine yields, finding ways to maximise both the number of cells that could be grown per millilitre of culture, and the number of virus particles that could be produced by each cell. Carina Joe, a post-doctoral researcher in Sandy's group, had figured out that by changing the growth medium (the nutritious broth in which the cells multiply) she could improve the yield up to fivefold (meaning that we would obtain five times more vaccine from a given volume of culture). They had so far, though, only tested this method at a small scale, in the research labs in Oxford. To prove that it worked well when scaled up would require access to a bioreactor as well as other specialised equipment. The best place to find these is in industrial bioprocessing plants, where they produce drugs like antibiotics or insulin. And for that we would need help. And more money.

Sarah had already put in her radioactive February application for £2 million from UKRI to make vaccines for, and run, our first clinical trials. But Sandy suggested that he could make a separate bid for funding to bring in an industrial partner to help us to develop the process at scale, so we would be ready for eventual manufacture. The aim was to very rapidly (within three months) prove that we had a process that could be used to make millions of doses. We would then be in the ideal position, as soon as we had evidence of safety and efficacy from clinical trials, of already knowing how to make enough vaccine to actually deploy it. We would also be in a much stronger position to find an industry partner for commercial manufacturing. We had made vaccines using this technology before, but never at scale, and neither AstraZeneca nor anyone else would have had much interest in an academic vaccine with no proven potential for large-scale manufacture.

The timing was excellent as I had been at a meeting of the UK BioIndustry Association earlier that week. It was mid-February when I had travelled down to London by train, sat tightly packed around a table in an airless meeting room, jostled elbow to elbow for a buffet lunch, and generally done a lot of mingling and chatting. During all the breakout sessions, I had been scoping out who in the UK would be interested in helping us with a big coronavirus project if one started to kick off. Afterwards I met my sister at an event she had organised at the Design Museum, then we went for dinner and cocktails on Kensington High Street to celebrate my birthday. It was a totally normal trip down to London – and of course at the time I didn't realise how long it would be until I would do anything like it again.

Although these kinds of conferences are sometimes seen as pointless, unproductive jollies, the relationships I had built there over years of soggy quiche slices and partially-frozen cheesecake turned out to be very important. It meant that when I came to these people with urgent and unprecedented requests in the coming weeks, there was great trust between us. So when I asked Netty at the BIA to put a call out explaining the work we already had in train, our confidence in our technology, and our need for assistance from industry, the response was heart-warming and incredible. We immediately received offers of support, equipment loans and expertise-sharing from across the UK. This enabled Sandy to rapidly finalise his request for funding. He asked for £400,000, to use his process to manufacture at fifty-litre and 200-litre scales at the Pall facility in Portsmouth, with support from the Vaccines Manufacturing and Innovation Centre team.* (The VMIC facility itself was, unfortunately, still in its construction phase in a field outside Didcot.)

The first test run at fifty-litre scale was started in Portsmouth in the last week of March, and a 200-litre run was started shortly afterwards. We were not, at this point, manufacturing vaccine that could be used in clinical trials. That would have required a rigorously controlled GMP-compliant facility. These were practice runs, testing methods, proving they could work at this scale, and getting an idea of yields. By mid-April, while we were quality-testing the first ever 700 vials of vaccine produced by the CBF, Pall was successfully producing fifty-litre and 200-litre scales of research-grade vaccine, and demonstrating excellent yields.† We had a route to large-scale manufacturing. The process was going to work.

---

* Sandy's bid was for a tiny amount of money for such a pivotal piece of work: he was able to make it so small because of the extraordinary goodwill of our partners, Pall in Portsmouth, Cobra in Keele and HALIX in the Netherlands, who were often contributing at minimal or no charge.

† Eventually they got to 2,000 doses per litre. This makes global supply possible: it means 2 million doses from a 1,000 litre run. By comparison, yields at the CBF and Advent were less than one hundred doses per litre. You could never manage to make enough for global supply from that.

It was a big moment. Up until then, we had an idea – a theory. We still didn't know if the idea would work: that was what the trials were for, to establish whether the vaccine was safe and effective. But even assuming it did work, our vaccine would remain only an idea, just an interesting piece of research with no impact in the real world, unless we found a way to make it at scale. A vaccine's ability to save lives is not only about a single efficacy number. It is also about how much you can make, how easily you can get it to people, and then, ultimately, how many people are willing to receive it. This was big because it was the moment we knew we were going to be able to make our vaccine – and potentially, if the vaccine worked, save lives – on a massive scale.

With this proven success, Sandy went back to the government with a request for more funding. He had developed a process and, refining it with Pall, proven it could work at scale. Now he needed more financing to contract with large-scale GMP-compliant manufacturers, so that they could stop doing whatever they had been planning to do in 2020 and get to work making our vaccine instead. This was getting really urgent. Sandy had pretty much arranged for large-scale manufacturing via a series of gentleman's agreements, but in the end the people being asked to deliver the goods are commercial companies, with costs and salaries to pay: there comes a point when goodwill runs out and they need a real contract. There were a tense few weeks of unproductive negotiations with possible funders. Then the cavalry arrived in the form of the brand-new Vaccines Taskforce. Combining technical expertise with a route to decision-making at the highest level of government, the Taskforce suddenly made what many had said was a crazy proposal in February a reality. Sandy was awarded funds to build a manufacturing consortium that could not only make lots of vaccine but also test it, store it, transport it and get it into vials (which requires highly specialist plant). This was not straightforward. Some manufacturers needed new equipment, or to repurpose existing equipment, some needed training, and everything was more difficult than normal because we were still in the middle of a pandemic.

One crucial new part of this set-up was Oxford BioMedica. Serendipitously they had a brand new and so far empty manufacturing suite just around the corner from us. Sandy suggested to the Taskforce that government funding could enable them to buy all the equipment needed to make our vaccine at scale, to guarantee a UK-based supply.* Sandy's funding was thus used to build a complete package to supply millions of doses to the UK, with the intention of being operational by the winter.†

By mid-April everything was in place to make our vaccine at scale: the production instructions or recipe, the starting material, the production cells, the funding, and a group of manufacturers to work with. Sandy and Adrian Hill had also brought the Serum Institute India (SII) on board. The SII are the world's biggest vaccine manufacturer. Their involvement would make it possible to produce enough vaccine not just for the UK but for the whole world, and irrespective of a country's ability to pay.

Then the AstraZeneca deal was announced on 30 April and things *really* took off. With their existing relationships with major manufacturing sites and the financing power to be able to commit to contracts, AstraZeneca was able to activate a programme of global production that was entirely beyond the scope of a UK

* The plan is that this equipment will eventually transfer to the VMIC and Oxford BioMedica can go back to their normal life.

† All these contracts then transferred from the government to AstraZeneca after they acquired the project.

university-led project. The EU took another three months to finalise their contract and start to transfer the manufacturing process. This disparity and nothing more nefarious than that is what lay behind the differences in the UK and EU supplies of vaccine by the time it was licensed in early 2021 that created such upset on the continent and led to talk of export bans. The UK had started manufacturing early – even before AstraZeneca came on board – and got months ahead with ironing out the inevitable glitches and teething problems.

On 5 June, AstraZeneca agreed to supply Covax, the global collaborative effort spearheaded by WHO, CEPI and Gavi* to help secure supply of vaccine for low- and middle-income countries, with up to 300 million doses.

Over the summer this network of large-scale manufacturers meant we were able to supply all the vaccine needed for our ever-expanding trials, now about to go global in Brazil and South Africa. By the end of October, AstraZeneca had set up manufacturing and supply agreements with more than twenty supply partners spread across the globe, including SII in India and SK Bio in South Korea, with the aim of supplying 3 billion doses of our vaccine – now officially renamed from the catchy ChAdOx1 nCoV-19 to the equally catchy AZD1222. Millions of them would be made by Oxford BioMedica, right on our doorstep. The UK had 100 million doses on order, the US had 300 million. AstraZeneca and Oxford were committed to providing the vaccine on a not-for-profit basis for the duration of the pandemic, and to low- and middle-income countries at no profit in perpetuity: a vaccine for the world.

* Gavi, The Vaccine Alliance, is an alliance of public and private sector organisations originally established to provide children living in lower-income counties with access to vaccines against deadly and debilitating diseases.

Between March and October, production had moved from our small team in Oxford, to a global network. We had gone from a first batch of 700 hand-crafted doses at the CBF, to 3,000 vials from Advent's first run in Italy, to millions of doses around the world.

Scale-up at this speed was an unprecedented achievement. We had a vaccine that we knew could be made in massive quantities, for a low price, and that could be stored in a fridge. And we had gone ahead and made massive quantities of it. But even now, in October, with all this money and all this effort committed, we still could not say for sure whether the vaccine actually worked.

CHAPTER 7

With Great Care, and Due Haste

1 January–30 December 2020
Confirmed cases: 0–82.83 million
Confirmed deaths: 0–1.81 million

Twenty years before the pandemic, when my children were about fifteen months old, I had a brief episode of complete amnesia. I was on my own in a car, driving towards a junction, and needed to get into either the left or the right lane. I realised that I didn't know which lane I wanted to be in because I had no idea where I was going. I also had no idea where I was, why I was in a car, or who I was. I was in full control of the vehicle, in a quiet suburban area, and driving within the speed limit, but I slowed down to give myself more time to work out what was going on. I told myself not to panic, but to think about what day it was. That seemed like a low-stress question to answer, rather than addressing who I was, or anything else challenging. Still nothing. Then, all of a sudden, everything came flooding back. It was a Saturday morning and I was on my way into the lab. I was on a road I had probably taken several thousand times before.

I was somewhat rattled by this episode, and took it as a sign that I probably needed less stress and more sleep. That was easier said than done. After the babies were born I had taken eighteen weeks of maternity leave, for eight of which the babies were in hospital. Then I had gone back to work full-time while my partner Rob took over at home. As a research scientist I was employed on a series of short-term contracts, typically three years long. Failure to deliver results would mean failure to secure more funding, and I would be out of a job. I couldn't afford to coast. On my first day back at work my supervisor had organised a two-hour meeting at 4 p.m. (this meant I was definitely going to miss bath time) to discuss the fact that he wanted to start clinical trials of the vaccine we were developing. This would be the first time the group had ventured into clinical testing, and I would be responsible for getting the necessary approvals, despite the fact that I had no experience in that field. The important fact I managed to latch onto was that we would be collaborating with a biotech company who had conducted trials before. I just about caught the name of the person there I should contact. The rest of that year is a blur.

Being a mother of triplets with a full-time job probably stood me in good stead for the challenges of 2020. Suddenly becoming the main breadwinner for a family of five, on a couple of hours' sleep a night: that was pressure. To be honest, although I was asked the question a lot, I rarely thought about how I was dealing with the pressure in 2020. Every day I simply had to deal with lots of different problems, each of which in itself was probably something I had dealt with before. There were more problems now, and shorter gaps between them. But I was working with a large team of skilled and hard-working people who were all experts in their field, all absolutely focused on doing the best job they possibly could, and all prepared to do whatever needed doing. Some very well-qualified people have taken a turn sticking bar-coded labels onto sample tubes.

The hard part was dealing with so many different parts of the process all at the same time, rather than in sequence. Starting from February, I was working through every weekend because that was the only time I was not constantly being interrupted. I would also wake up in the early hours thinking about a job that needed to be done, or a question asked or answered. Then I would think of a second, and a third, and before I knew it I would be heading downstairs to my computer before I forgot all the details of what had suddenly become clear to me. I never set an alarm or forced myself out of bed, but a 4 a.m. start became normal for a while. It wasn't the first time I had managed on too little sleep; nor the first time I had had to cut out from my life anything that wasn't strictly necessary. Plus, as we went into successive lockdowns there was not much else I *could* do but work.

In March and April, I would wake up worried that the whole programme would be derailed because between us we had managed to overlook something important. The team had more or less assembled itself from the few of us who had started the project, others we asked to join, volunteers, and people whose own research had stopped because of the pandemic. It had been a handful of people in January and was around 300 by July – not including the colleagues we were working with at Advent, AstraZeneca, in Brazil and South Africa, at the MHRA and so on. Groups of people were working together who had not worked together before, which added to my sense that something no one had thought about could come along and trip us up. But I was not the only one constantly thinking things through, trying to anticipate problems, and looking for things we might have missed. Everything got done.

Later in the year, any time there was a lull, my adrenaline levels dropped and exhaustion set in. It happened in August, when I nearly fell asleep at the kitchen table over dinner one evening, then slept for twelve hours straight. Rob was worried that the fatigue meant I might have Covid (I didn't). It happened again in the middle of December, when the efficacy results had been published but we were still waiting for the second shoe to drop, which would be when the MHRA announced emergency-use licensure of the vaccine. I slept like the dead for one night.

Every member of the team working harder than they ever had before was part of what made it possible to develop the vaccine in record time, but of course it was not the whole story. As often as I was asked how I was dealing with the pressure, an even more common question was how it could be possible to turn what was usually a ten-year process into a one-year process without cutting any corners. Some people were relieved that we were making such quick progress, but for many the speed with which we were working led to worries and mistrust. There was a concern that if we were doing it so fast we must be 'rushing it': taking less care or missing out important parts of the process. The answer to that question is really the thread running through this whole book, but in this chapter I want to summarise it as clearly as I can.

It is true that vaccines have in the past taken a long, long time to develop. Until 2020, a new vaccine usually took at least ten years to develop from concept to roll-out. Many took much longer. The malaria vaccine programme at the Jenner Institute has been going for twenty-five years, and research into malaria vaccines had been going on for more than a hundred years – so far, with limited success. The lab-to-jab record-holder was the mumps vaccine, developed in four years by Maurice Hilleman in the United States in the 1960s.[1]

But the standard lengthy timeline we were all used to was never because vaccine development required ten, fifteen or thirty years of continuous painstaking lab work, clinical trials and data analysis. For every vaccine that had ever been developed up until 2020, most of the elapsed development time was spent waiting. In 2020, there were three key factors that enabled us to cut out the waiting and crunch ten years into one: first, the work we had already done; second, changes to the way funding was given out; and third, doing in parallel things that we would normally do in sequence.

In some ways, it is misleading to say that we developed the vaccine in a year. A large part of the reason we were able to move so fast in 2020 was due to all of the work we had been doing, both on other vaccines and on planning for Disease X, in the years leading up to it. We were, unfortunately, not as well prepared as we could have been. For example, we had not been able to do as much work as we would have liked on speeding up the early, DNA-sequence-to-vial part of our development process (including my new 'rapid method' to make the starting materials and Sandy's new purification method). So we did not have the absolutely best route fully mapped out. Our plan was not finalised. But we were also not starting from scratch. We had years of experience and a large body of work behind us. Thanks to that, we had a good idea of where we were going and how best to get there.

Most importantly, we did not need to design, make and test a brand-new vaccine. We had a tried-and-tested platform technology, that we had been working on for years – starting with a flu vaccine back in 2012 and including a vaccine against MERS, another coronavirus. That meant that before we even knew the pathogen's genome, we knew the design for our vaccine – the gene coding for the SARS-CoV-2 spike protein, plugged into ChAdOx1 – and once we did receive the genome, we were able to design the exact DNA sequence we needed within forty-eight hours. Less than four months after that we had the first doses made, quality-assured, and ready to use in clinical trials. Because of the many safety trials we had done on previous adenovirus-vectored vaccines, with subjects aged from 1 to 90, we then had a large amount of data on what dosage would be safe and what dosage would induce the best immune response.

All of those international conferences also came good. It meant that we had a huge network of trusted colleagues all over the world. When we wanted to set up trials in Brazil or South Africa, Andy wasn't just pitching up unannounced at a hospital in Rio or Johannesburg. We had existing relationships, we knew all the people already, and we knew they would be able to conduct high-quality clinical trials because they already had the infrastructure and experience.

The 'plug and play' approach of a platform technology – a safe, effective, proven delivery system that can be quickly adapted to a new disease – was not available to Maurice Hilleman in the 1960s. He spent two years getting to the stage we reached in four months, transferring a virus taken from his daughter's throat through a succession of chicken eggs (a technique that had been used to produce other viral vaccines) until he judged it had been weakened enough to be trialled as a vaccine.

But, whilst technological advances were crucial to our ability to work quickly in 2020, they were not the only or even the main factor at play. Decades after Hilleman's work on mumps, despite all the advances in technology, getting from idea to first clinical trials in two years would still be seen as extremely fast, and even five years was considered very efficient. What had enabled Maurice Hilleman to move so quickly compared with those who came after him was that he was in the fortunate position of having access to the support and funding he needed, when he needed it.

Access to funding was the second reason we were able to move so quickly in 2020. The truth is that previous vaccine development – including for diseases of significant global burden like malaria, and against pathogens that could wreak as much havoc as Covid has done, like influenza A (which includes avian flu) – had been slow not because it was impossible or unsafe to go fast, but because going fast was not seen as a high enough priority. Developing a vaccine is expensive. Manufacturing a vaccine is even more expensive. For every vaccine development programme, there has to be, first, a reason for someone to start, and second, ever-increasing levels of funding to keep going. Small companies and university scientists are those most likely to start a piece of research, but tend not to have the skill sets or facilities to take things all the way through to licensure. Large pharmaceutical companies, which are required to make a profit for their shareholders, might have the skill sets and facilities needed to get a vaccine licensed and to manufacture at scale, but not the motivation. For example, if a vaccine is likely to only ever be needed in small quantities (like the vaccines for Nipah or Lassa fever or MERS – and remember the Ebola vaccine tested in the 2014 outbreak was originally developed as part of a US bio-defence programme, not to protect people in West Africa), there's not much commercial incentive. Similarly, if a new vaccine candidate works well from an immunological point of view, but would be too expensive, or impractical in some other way, to manufacture at scale, no one is likely to make it.

An example of an academic-led project that did not make it beyond phase I trials for reasons of impracticality was a Covid-19 vaccine project from the University of Queensland in Australia. A protein used in the vaccine induced those vaccinated to produce antibodies to a part of the HIV virus. This meant that if they then took an HIV test, it would appear to be positive. There was no danger that the trial participants had been infected with HIV, but since there was also no chance that the HIV test would be changed to allow the vaccine to be used without interfering with the test, vaccine development had to stop.[2]

Within this landscape, research funding bodies can be reluctant to support development work beyond the earliest, least costly, stages, when they know that without the support of a large commercial partner it is unlikely to reach completion. This is the 'market failure' that organisations like CEPI and Gavi and the Gates Foundation have been trying to address. We call it the 'funding valley of death' because so many hopeful projects fall into it, never to be seen again. It is an issue we had been trying to address at Oxford University, where we took the view that we needed to be able not only to research a vaccine candidate in the lab, but also to manufacture it for clinical trials, and conduct at least the phase I safety trials ourselves. However, even with the right skill sets and facilities, we still had to find the funding to pay for our work.

In normal circumstances, that funding comes in tranches. You apply for a first tranche of funding, and wait to hear whether you have been successful. That alone can easily take a year. Then you do a bit of work, perhaps demonstrating that the vaccine is capable of inducing an immune response in animals and protecting animals against the pathogen. You publish the results, and present your work at conferences to raise its profile and attract the attention of funders. That maybe takes another year or two. Then, if your results look promising, you apply for the next, much larger tranche of funding, to manufacture the vaccine in readiness for clinical trials, apply for approvals and vaccinate a small number of people in a phase I study. You wait to hear whether you have been successful, which can easily take a year. Perhaps this time you aren't successful. So you write another application, wait another year. And so on.

It is as if you are making a roast dinner and for every ingredient you have to make a separate trip to the shops to buy it, then cook it and demonstrate that it is going to be delicious, before moving on to the next. And even for us, with the CBF on site and the ability to run clinical trials, once we got to the stage of wanting to run large-scale phase III trials, it went beyond the kind of funding that was available to a university. We had to publish our studies and wait for interest from pharmaceutical companies or large-scale endeavours like CEPI. That was like getting to the point of having successfully cooked all the vegetables, then having to just put them, steaming, in the front window and hope that someone would offer to sell you a chicken.

In 2020, this kind of limping along, with a delay of several years between deciding to start making a vaccine and the first clinical trials, simply would not do. The pandemic was killing hundreds of thousands of people and shutting down entire societies and it became imperative and urgent to have a vaccine to control it. In January, February and March we were out on a limb, working at financial risk to the university, and probably at even greater risk to our own professional reputations, but by April there was widespread recognition that resourcing research into Covid-19 vaccines was an urgent priority. We were allowed to do a big shop and put all the ingredients we needed in the trolley all at once.

In 2020 usual funding cycles were enormously compressed, cutting literally years out of the development process. The usual amounts of money available were also hugely increased. By the end of April, having been operating on less than a shoestring, we had £20 million of government funding and access to the vaccine development and large-scale manufacturing capabilities of AstraZeneca. In May the US government put another $1.2 billion into the US trials for the Oxford AstraZeneca vaccine through its Operation Warp Speed programme.[3]

As vaccine development was transformed from a funding Cinderella into royalty, other crucial resources flowed just as abundantly. Cath's CBF dropped its work on the Ebola vaccine to prioritise making the Covid vaccine. A process that would normally take months – finding the hundreds of volunteers we needed for our first clinical trials – was completed in hours because people were so keen to help. The MHRA prioritised the work they needed to do to review and approve first our trials and later our results, putting us and other Covid vaccines at the front of the queue, bringing in dozens of outside experts to help, and, like us, working long hours, seven days a week. As other work was deprioritised we gained access to a huge pool of skilled scientists who would usually have been working on other trials or research. And scientists and other experts from around the world were more willing than ever before to co-operate and collaborate. I joined calls led by the WHO in which many groups developing vaccines presented their plans and reported on progress. We all wanted to learn from each other, and knew that, unlike normal, this was not a race or a competition for a limited market where the winner would take all. For some commercial vaccines, being first means making the most money, because it is difficult for other companies to break into the market once the first vaccine is in widespread use. With 7 billion people at risk, it was going to take lots of vaccines made by different companies in different countries and in different ways to beat the virus, and we needed to share our knowledge to get there as fast as possible.

The third reason we were able to move so quickly was that from the start we proceeded 'at risk' – doing in parallel and back-to-back things that would usually be done in sequence with long pauses in between. As we've explained, 'at risk' did not mean at risk to the safety of the vaccine. The risk was to those of us working on the vaccine, that we would have wasted our time and money. So, for example, usually we would not start work on making the clinical-grade vaccine in Cath's CBF until we had shown the research-grade vaccine worked in animals. This time, we did those two pieces of work in parallel, the risk being that if the vaccine did not work in animals we would have wasted our efforts manufacturing it. Similarly, usually we would not start work on the design of a clinical trial until we had completed all the preclinical (animal) trials. That way, if the preclinical trials showed that the vaccine was not safe, or not effective, we would not have wasted anyone's time preparing clinical trials that could not go ahead. This time, we did all the design and preparation of the clinical trials, including recruiting and screening the volunteers, whilst the preclinical trials were still going on. That way, it was literally the day after we had the safety data from our preclinical trials, which was also the day that Cath's CBF had vials of the vaccine ready to use, that we were putting the vaccine into the arms of our first volunteers in phase I safety trials.

We overlapped phases of clinical trials, once we had the data we needed on safety. We started scaling up manufacturing before we had even done our first clinical trials. You wouldn't dream of doing that in any normal world because it is so expensive. AstraZeneca and its partners manufactured millions of doses of vaccine before we had data showing that the vaccine worked: that was even more expensive. It meant that if the vaccine turned out not to work they would have to throw millions of doses away. We started talking to the MHRA early, we kept talking to them all the way through, and they reviewed all of our data – over 500,000 pages of it – on a rolling basis rather than waiting to start until we had provided every last piece of evidence. Under the rolling review, the MHRA looked at everything they always look at: the preclinical data, the manufacturing data and the safety and efficacy data. And they looked at it just as carefully as they always do. But they started sooner, and put more people on it. That is how they got it done quicker.

The last piece of waiting that got cut short, of course, was the wait for the results of our efficacy trials. In 2014 during the Ebola outbreak, by the time the phase III trial started the outbreak was being brought under control. This made it difficult to assess whether the vaccine being tested was effective or not as people on the trial were not being exposed to the virus. But in 2020, with the Covid-19 pandemic spreading so fast around the world, and with our volunteers mostly healthcare workers who were more likely than most to be exposed to the virus, we would not have so long to wait.

Before 2020, no one had ever developed a vaccine in a year. But that was not because it could not be done. It was because it had never been tried. We were able to go faster in 2020 not because we cut any corners or took risks with our product. We still did every single thing that needed to be done to develop a vaccine safely. We did not miss out any steps. Nor was any individual task – filling a vial, vaccinating a volunteer, analysing a graph – done with less than the usual care and attention. We went faster because we had to this time – the world needed the vaccine as soon as possible and, as we know from seeing the daily death rates, every day counts.

CHAPTER 8

Trials

23 April–20 July 2020
Confirmed cases: 2.74 million–14.72 million
Confirmed deaths: 197,150–606,899

Thursday 23 April was a memorable day for me on many levels. I woke up feeling really nervous. Not because it was the day we were going to put the first shot of our vaccine into the arm of our first volunteer in our first trial: I had every confidence that that would go smoothly. But because I was scheduled to do a radio interview with LBC's James O'Brien, and he's a tough interviewer.

I'm completely happy standing on stage and talking science to an audience of scientists, but this felt very different. I knew that James would ask broad questions, and I'm not a vaccinologist or a medical doctor, and I didn't want to let anyone down by saying anything wrong or stupid or that could be taken the wrong way. I was having a bad case of imposter syndrome. In the end, it seemed to go pretty well. I think I cheered James up. Everything was feeling a bit gloomy at this point: we had been locked down for a month, the prime minister had still not returned to work having been in intensive care, PPE shortages were so bad that some doctors and nurses were being asked to work in bin bags, the UN had warned that Covid-19 could cause famines of 'biblical proportions', and the chief medical officer, Chris Whitty, had used the previous day's government coronavirus briefing to say that it was 'wholly unrealistic' to expect life to return to normal soon. He also said that the way out would be a vaccine or treatment, but that it was extremely unlikely we would find either before the end of the calendar year.[1] But my message was one of hope: the trials have started; we are confident that this vaccine will work; we will find a way out of this terrible situation.

While I was being self-indulgent worrying about my radio interview, the real action was up the hill at the university's Centre for Clinical Vaccinology and Tropical Medicine (we call it the CCVTM, admittedly only a slight improvement). This is the small facility, across the road from mine, where volunteers come to be vaccinated in our clinical trials. Today we were vaccinating just two people – Elisa Granato, a microbiologist, and Edward O'Neill, a cancer researcher. After taking blood samples from each of them, the nurse injected one with the real vaccine and one with a placebo. Neither they nor the nurse knew who was getting which. Afterwards, our volunteers were monitored in the clinic for an hour, for any unexpected side effects such as allergic reactions. Then they were free to go. Neither Sarah nor I were there to witness this moment – we were being very strict about social distancing – but Fergus Walsh, the BBC's medical editor, was. So like a lot of other people I saw it that evening on the news. It triggered a mix of emotions: relief and optimism, gratitude and nerves. I am used to making vaccines that get used in clinical trials – that's my job. But usually the vaccine in question is unlikely to have any direct impact on me or my loved ones. This time it was personal.

—

It took us sixty-five days, between January and April, to make our first batch of vaccine. It took us another seven months, from April to November, to test it, in a series of ever-expanding clinical trials.

Most of the work, including working through the unthinkably huge amounts of logistical details required to get a trial up and running so quickly during a pandemic, was done by the Oxford Vaccine Group, which is Andy's team. Over 200 people from Oxford worked on the clinical trials, each playing their own part. Andy was the conductor, with various soloists taking turns to make their contributions, but everyone, including the second violins in the back row, had a vital part to play.

Clinical trials of vaccines are divided into three phases. There are no hard and fast rules about how many people should be included in each phase, or how long they should last.

Phase I trials, also called first-in-human trials, test for safety. Because there is always a risk that a new medicine won't behave as expected, they start small. They will also always involve healthy adults aged between 18 and 50 or 55. These volunteers are screened before vaccination to make sure that they are indeed healthy. After vaccination, their health will be intensely scrutinised. Blood samples will be taken, and checked for anything abnormal. The volunteers also have a health diary to fill in. Every day for a week they are asked to say which 'solicited adverse events' they are experiencing. 'Adverse events' in this context means side effects. A solicited adverse event is one that we are expecting, and have asked about, such as a sore arm or a headache. We also ask about unsolicited adverse events for the full duration of the trial, typically a year, meaning anything else bad that happens healthwise, even if the volunteer does not think it has anything to do with the vaccination. We then assess each event to determine whether it is definitely, probably, possibly, or not related to vaccination. As you can imagine, we end up with a varied list of 'adverse events'. In Oxford, bike accidents crop up quite a lot. There was a trial in Baltimore where gunshot wounds were recorded more than once. But we have to collect the information to make sure we don't miss something important. If there is an unexpected increase in any type of adverse event, it could be a consequence of the vaccine. And if there is a concern that the vaccine is causing unexpected or serious adverse events, the trial will stop while we investigate.

The reason we start trials in healthy young adults first is not so much that vaccines are more dangerous to unhealthy or older people (although it does make sense to test a new medicine on the strongest members of the population); it is also that unhealthy or older people are more dangerous to vaccine trials. Unfortunately, people with underlying health conditions and older people are more likely to experience poor health, and if one of the first volunteers becomes ill soon after vaccination, it is more difficult to know whether the vaccination could have been responsible, and so more likely that we will have to stop the trial to investigate.

Although the primary aim of a phase I vaccine trial is to look at safety, immune responses will be looked at too. All vaccines work by inducing the body to produce an immune response so there is no point continuing with a vaccine that does not induce an immune response, no matter how safe it is.

Once a phase I trial has demonstrated acceptable safety and shown that immune responses are induced in young healthy adults, a phase II trial can include more people with a wider age range. Again, we test for safety and for immune response. As people age, the immune system doesn't respond to new information as well as it used to. As a result, some vaccines don't produce such a strong immune response in older people.

That doesn't mean there is nothing we can do about it: we are learning all the time how to design vaccines that do induce a good response in older people.* But this is why it is important to test new vaccines in different age groups. We need to know which ones are going to be effective in older people, and which ones aren't, and then use the vaccines accordingly. In phase II, volunteers again undergo health screening and we only recruit those without pre-existing conditions, so that we are only looking at changes in response to the vaccine due to age, rather than age plus diabetes or cardiovascular disorders, for example.

The phase III trial will include a much larger number of people, this time with no pre-screening for underlying health conditions. It is in phase III that information about whether the vaccine protects people against the disease is collected. In phases I and II we look at blood samples to see whether volunteers are producing good immune responses, but in phase III we look at whether those immune responses are actually protective against disease. Do people who have been vaccinated fall ill when exposed to the pathogen, or not?

Typically, all three phases of a vaccine trial will be blinded randomised placebo-controlled trials. That means that half the volunteers receive the real vaccine and the other half receive a placebo. The volunteers don't know which one they are receiving and nor do the people administering the vaccines, processing the records of adverse events or carrying out the immunology tests. By eventually comparing the data from the vaccine group and the placebo group, we can draw conclusions about the effects of the vaccine.

---

* There are now influenza vaccines, for example, aimed specifically at older people, which use either a higher dose or extra ingredients to improve the immune response. A higher dose that might result in a fever in young people might be the right dose to give to an older person.

To know whether a vaccine protects people from disease, some people in the trial have to become infected with that disease (which can be problematic, as Sarah illustrated in Chapter 2 in the case of Ebola). When enough people have tested positive, we can 'unblind' the trial, meaning we look at which vaccine those people received. If all the infections are in the placebo group, the vaccine is extremely effective. If equal numbers have received the vaccine and the placebo, then the vaccine is not effective. In real life, the number will be somewhere in between.

In normal times it takes several years to progress from phase I to phase III trials. But of course, Covid was different. As in so many other areas of this vaccine's development, the trials team had to do all the work they would usually do, and collect all the safety data they would usually collect – and more – but without the usual long pauses between stages. In fact, the stages overlapped.

This meant that I learned a lot. Usually, by the time one of our vaccines goes to clinical trials, my team has moved onto another project. Not in this case. In this case, we were still making the vaccine while the trials were being planned and carried out. Running the trials at this pace also meant that the team running the vaccinations and the immunology teams analysing blood samples taken from our volunteers had to work phenomenally hard. I can't overstate their extraordinary commitment. As with so many of the milestones on this project, 23 April 2020, when our vaccine was given to a person for the first time, was both the culmination of a huge amount of work, and the beginning of a lot more.

—

Designing clinical trials means first of all answering a lot of questions: who will our volunteers be? How old? How many? What dose will we give them? One dose or two? How far apart? What will we measure during the trial? How long will the trial run for?

In 2020, many of these decisions were relatively straightforward. In other cases, we were having to make difficult calls in a fast-moving situation where we necessarily did not have full information. Some of these decisions would turn out to have significant consequences.

For example, as is usual, we started off testing our vaccine on young, healthy adults. We then widened the age range, first adding 56- to 69-year-olds, and then over 70s. The decision to delay bringing older people into the trial until there was a lot of safety data was cautious and would normally have been uncontroversial, but it caused us some problems down the line. Together with the dramatic effect of the UK-wide lockdown on disease transmission, it meant that by the time older people had been fully vaccinated in our trial, case numbers were very low. Also, older people were behaving very cautiously – meaning they were not going out and getting infected. As a result, very few volunteers over the age of 65 had tested positive for Covid by the time the trial reported its results on 23 November. With hindsight it is easy to say that we should have included older people in the trial earlier, so that we had more definitive data, sooner, on whether the vaccine protected older people. But at the time we designed the trial it seemed right, from a safety point of view, to take the standard approach of starting with younger, healthy adults and expanding from there.

We also thought a lot about whether we would test one dose or two. The advantages of a one-dose vaccine are obvious. In a pandemic, we want to see rapid protection after vaccination without having to wait until a second dose is given. We did start out planning a one-dose vaccine. At this point we thought we were making an 'outbreak' vaccine to contain the spread of disease as quickly as possible, and we knew from previous trials of other vaccines on the platform that ChAdOx1 would result in strong immune responses after just one dose. However, we also gave a small number of people in our phase I trial two doses, four weeks apart. Why? Because we are scientists and so our instinct is to try to find things out. And because there was so much uncertainty at that point, it made sense to keep our options open.

By the time we started to get data suggesting that those who had received two doses were producing a better immune response, it was becoming clear that Covid-19 was spreading very fast around the whole world; could possibly reinfect people who had already been infected; and was likely to continue to circulate for many years to come. In these circumstances, it was possible that having a highly effective vaccine would be more important than having a fast-acting one. We knew that one dose was inducing an immune response. But we could also see that the immune response from two doses was better. We didn't know at this stage how strong the immune response needed to be to protect against infection, so rather than give everybody in the trial one dose and then discover that efficacy was lower than we had hoped for, we opted to change the trial plan to give everyone two doses.

One benefit of the fact that our vaccine was based on a platform that had been used in multiple previous trials was that we did not have to do a 'dose escalation' study. The first vaccination with ChAdOx1, in 2012, was in an influenza vaccine trial. Before this, there had been many vaccine trials of other replication-deficient adenoviral-vectored vaccines. There was no reason to expect that the safety profile (the reactions experienced by people being vaccinated) would be any different with ChAdOx1, but we still proceeded with caution. In this 2012 flu vaccine trial, the first three people to be vaccinated received a dose one hundred times lower than the dose we anticipated using in later trials. This was not expected to result in much of an immune response but was just to make sure there was no unforeseen safety problem. Rather than vaccinate all three on the same day, only one person was vaccinated first. They stayed in the clinic for an hour to make sure there was no immediate allergic reaction, received a follow-up phone call from the trial doctor the next day to check that all was well, and then came back to the clinic for a check-up the day after that. As there were no problems, two more people were vaccinated at the same dose and all of the safety data from the participant diaries and check-ups two weeks later was reviewed before we were given permission to move on. The next three volunteers were given a dose ten times higher – so still ten times lower than we would expect to continue using – and the same procedure of vaccinating first one and then two more people was followed. A third group was given a dose five times higher, and a fourth group, finally, was given the full dose.

Having been through that process with the flu vaccine, later trials of ChAdOx1 did not need to start from such a low dose, and for the MERS vaccine trial in 2018 three different doses had been tested, omitting the very low one used in the influenza vaccine trial. For our Covid vaccine, ChAdOx1 nCoV-19, having tested multiple ChAdOx1 vaccines previously, and knowing that we needed rapid onset of an immune response in a pandemic, only the highest dose was tested.

We also knew that, in order to assess the vaccine's ability to protect people from disease, we would have to wait for our volunteers to catch Covid-19 in the course of their normal lives. With some diseases, such as malaria or influenza, we can conduct what we call challenge trials, in which volunteers are intentionally exposed to the pathogen. This has the advantage of yielding results very quickly. However, at the time we were designing the trials for our Covid-19 vaccine, relatively little was known about SARS-CoV-2 (the pathogen) or Covid-19 (the disease). The ethical implications of exposing people to a dangerous and poorly studied virus with no known treatment meant that challenge trials were, for the time being, off the table.* When we were first planning the vaccine trials, cases in the UK were increasing and predicted to peak in May. We thought that if we could get around 1,000 people immunised by then, we should be able to see if the vaccine was working by as early as June.

But of course, cases in the UK increased faster in February and March than anyone had expected. First a university student and his mother were diagnosed, quarantined and recovered without anyone else becoming infected. Then there was a family who had been on a ski holiday in Europe. And then suddenly, there were lots of cases and no understanding of where they had come from. The national lockdown reduced the daily number of new cases, which was crucial for protection of the population as a whole, but problematic for our trial. With fewer people becoming infected, we would still get our safety and immune response data on the same timetable, but we would have to wait longer for the data on efficacy.

As well as designing the trial and getting approvals, the clinics had to be prepared. Nurses needed to be trained to administer

* The UK regulators did in fact approve the world's first coronavirus human challenge trials in February 2021.

the vaccine. Information leaflets and consent forms had to be drafted. And most importantly, our first 1,000 volunteers needed to be recruited and screened. Trials also have to be registered in the public domain (ClinicalTrials.gov), to ensure that no one can start a trial and then not report the findings if the results are not as good as hoped. All this preparation, which of course would usually only start once funding has been secured, would normally take at least six months. When Sarah ran her phase I clinical trial for the MERS vaccine in 2018, she needed twenty-four volunteers and it took thirty-one weeks to recruit and screen them.

Another critical piece of the puzzle that was needed before we could start in-human trials was the results of our preclinical studies that were being performed in macaque monkeys. Although Sarah (and others) had trialled ChAdOx1-vectored vaccines in people previously, this precise vaccine, with the SARS-CoV-2 spike, was of course brand new.

From a very early stage, Richard Cornall, the head of the university's Nuffield Department of Medicine, had instituted weekly meetings of everyone working on Covid-19-related research projects. The idea was to discuss findings, share progress, and bring our knowledge to bear on each other's work. This was extremely unusual – some of these academics were more used to competing than collaborating – and I thought Richard's initiative showed great foresight. Every week the Zoom call was full of up to fifty people representing the teams developing the Covid testing, developing the Covid track and trace app, running the RECOVERY trial testing potential Covid treatments, looking at the immunology in infected people, understanding the virus protein structure, making antibodies, doing the genome sequencing, trying to set up challenge studies in animals, and so on. I am sure that this is part of the reason that the University of Oxford led on so much Covid-19-related research that year.*

Some vaccines have in the past caused people to suffer worse disease if they become infected naturally later on. This had happened with a vaccine against respiratory syncytial virus in the 1960s. Although there were no problems immediately after vaccination, which is when we would usually expect to see any adverse events, some of the children who were vaccinated developed very severe disease when they later encountered the virus. Something similar appeared to be happening with some attempts to make a vaccine against SARS, although the vaccines in question were only tested in animals. The SARS outbreak was contained without vaccines, but even if it had not been, a vaccine that was possibly causing antibody-induced enhancement in animal trials would not have been a good candidate for further development. A lot has been learned about vaccines and the immune system since the 1960s and we now understand what type of vaccine has the potential to cause this kind of vaccine-enhanced disease, and what type of immune response is involved.

The only way to test this before first-in-human trials was to vaccinate and then deliberately expose animals to very high doses of the coronavirus. We thought a lot about whether we should delay the start of our human trials and wait for the results of these animal tests. We knew that every day counted, especially now that we were in lockdown. Because of falling case numbers, starting even a few days later could delay our results by weeks. Sarah and Tess had a high degree of confidence that they understood this issue well and that it would not be a problem. No vaccine developed on the ChAdOx1 platform had ever caused

* In March 2021, the NHS reported that the inexpensive and widely available steroid dexamethasone, identified as a treatment for Covid-19 through the RECOVERY trial, had saved 22,000 lives in the UK and an estimated 1 million lives worldwide.

this kind of response, including the very similar ChAdOx1 MERS vaccine which had been tested in a clinical trial. On the other hand, the strong view being pushed by some at Richard Cornall's weekly meeting was that waiting would provide an additional layer of confidence in the safety of the vaccine. It was another judgement call and because it was about safety, we erred on the side of caution.

The studies were carried out by our collaborators at the National Institutes of Health Rocky Mountain Labs in the United States. We were receiving frequent updates and had most of the results by the afternoon of the 21st, with the final email into Sarah's inbox at 3.28 a.m. on the morning of the 23rd. The results were as we had expected based on all of our other studies. The vaccine induced the right kind of immune response, not the wrong kind. It didn't cause any worsening of the effects of the virus. In fact, the vaccinated animals did better after exposure to these massive doses of coronavirus than the control animals. This was already strong evidence that the vaccine would prove to be both safe and effective.

—

So, the trials – and a whole new set of challenges – could begin. Phase I trials might typically include a few dozen people, but our plan was to vaccinate about 1,000 people: 500 with a single dose of our vaccine, and 500 with a single dose of the placebo.* Again we started gradually: two people on the first day, six on

* We used the meningitis vaccine MenACWY as a placebo, rather than saline. This was because, unlike saline solution, it has similar minor side effects to the real vaccine – sore arm, headache, slight fever – making it harder for volunteers to guess whether they had received the real vaccine or not, and therefore making the trial better blinded and better controlled.

the third day, then quickly ramping up from there until we were vaccinating around a hundred people a day. There was also a small group of ten volunteers who were given a second, booster dose, twenty-eight days after the first dose, so that the team could assess whether that improved their immune response.

The start of our clinical trials made us, suddenly, extremely public property. I was on LBC, Andy was on BBC *Breakfast* and Sarah was on the *Andrew Marr Show*. Three days after we vaccinated our first two volunteers, a false claim that Elisa Granato had died went viral on social media. The completely untrue article – really, who on earth would use their time to make up something like that? – looked credible: it was understandable that people were taken in by it, and it offered a really good insight into how easy it is to spread fake news. It was unpleasant for Elisa and her family, and I am sure very worrying for the other volunteers signed up for the trial, but Elisa took it in her stride, posting on Twitter and then doing an online interview with Fergus Walsh for the BBC.* People had posted on her Twitter feed from all over the world saying that they would not believe she was still alive until they saw her in a date-stamped interview. The Department of Health put out a statement too. I wonder what the person who created the fake story got out of it.

That same weekend I was at a Zoom party (hopefully in a couple of years we will all have completely forgotten what that phrase means). Mid-Zoom, I got a weird call from the university Internet security team saying that they had noticed unusual activity on my Twitter account and asking whether I was OK:

* His tweet read: 'Fake news has been circulating on social media that the first volunteer in the Oxford vaccine trial has died. This is not true! I spent several minutes this morning chatting with Elisa Granato via Skype. She is very much alive and told me she is feeling "absolutely fine".'

was I feeling worried or threatened? I wasn't, because I don't really check Twitter that religiously. I had posted a thank you to my team on the day of the first vaccination, naming them all in an attempt to recognise each person's efforts and contribution that had got us to this point. I was becoming conscious, I suppose, that as with all team efforts, there would be people who ended up very visible and people who disappeared into the background. The problem was, as the security team now gently explained to me, I wasn't supposed to name people.* Also, my tweet had got people going. A lot of people had liked it, tweeting thanks and congratulations, which was lovely. But it had also attracted some committed anti-vaxxers, and then there was a small Twitter-off as someone tried to call it a British success story and other people jumped in pointing out that it was an international team.

To me this is, like most things, more complicated than a tweet really allowed for. I am proud of the international nature of my team, and of UK science in general. Diversity of ideas helps progress. At the same time, we all live and work in the UK by choice, and although I will not be happy if the current government tries to take the credit, there is a lot about this country (the flexibility and creativity of the university, the cooperation of the bioscience industry, the backbone of the NHS) that did contribute to the success of the vaccine, and I am proud of that.

However, after a whole weekend was completely lost to dealing with the media in one form or another, we decided we needed a media blackout. The attention was all starting to get a bit intense and personal, and very distracting. We were

---

* I had assumed that we were all easy to find anyway through the CBF's website, but it turned out that wasn't possible because the website was so out of date. I still haven't had time to update it.

vaccinating members of the public now and just needed to focus on getting it done and getting it done properly.

Our volunteers filled in daily online health diaries for the first week after vaccination, and we also asked them to record any other health issues they were experiencing for a month, and any serious events for a year. This was so that we could assess the vaccine's safety. They also came back to the clinic at intervals (after 3, 7, 14, 28 and 56 days) for check-ups and to give blood samples so that we could study their immune response.* This was where the immunologists came in.

The cells of your immune system circulate in your blood – these are your white blood cells. As discussed in Chapter 3, this army of cells is your defence mechanism against pathogens, and for the purpose of understanding vaccines there are two types that are the most important: B cells and T cells. B cells make the antibodies that will help to fight off future infections, and T cells can recognise infected cells that have been turned into virus factories, and target them for destruction before they can finish the job. We already knew that one of the key benefits of the ChAdOx1 platform was that these types of vaccines generated both B cell and T cell responses, giving your immune system a two-pronged response to fight infection. We wanted to know whether the ChAdOx1 nCoV-19 vaccine was behaving as we expected it to and generating both antibodies and T cells. The blood tests would tell us that. Each volunteer donated about three tablespoons of blood at each visit, which was split into a number of small tubes, ready for a number of different tests. One test looked at whether the blood sample contained antibodies able to bind to parts of the SARS-CoV-2 virus. A second

* They also came back after 182 and 364 days. Some volunteers who were recruited into the trial later were only asked to come back to give blood samples at days 28, 182 and 364.

asked whether the antibodies could neutralise the virus and prevent it from entering human cells. And a third looked for the presence of T cells.

Even at this first phase I stage, the team in the immunology lab was working all hours as samples arrived from day, evening and weekend clinics. And not just round the clock but against the clock too: samples had to be processed within four hours or the cells would start to die and the tests would be void. People were working in the lab until two or three in the morning, and up to ten days back-to-back. Two couples made donations for us to have healthy meals delivered to the site every day for months. It made us feel noticed and supported and more importantly it stopped us feeling hungry. The catering facilities were shut, and we were all running out of change for the vending machines, and hungry people make mistakes.* We also received lots of colourful pictures from children that we put up on the walls: drawings of the virus, and their suggestions of how to beat it and thanks for what we were doing, which were very motivating when we were tired and stressed.† Morale was high though. There was a sense of purpose and excitement, and enormous goodwill because everyone on the team knew their work was making a difference.

As well as the diaries and the blood tests, the participants were also reporting any Covid-19 symptoms, and those with symptoms were immediately tested for the presence of the virus and given medical help if needed. If the test came back positive, it might count as a 'case' that would be part of the calculation on whether the vaccine was working or not, although we also

* Thank you Jonathan and Tracy Turner, and Denise Foderato and Frank Quattrone – this was a godsend and you are heroes.

† I am going to make sure they are all kept and donated to the History of Science Museum in Oxford, where they are archiving the trial materials.

had to take into account how long it was since they had been vaccinated.

All the phase I vaccinations were done by 21 May. On 22 May we started recruiting for our combined phase II/III study, and the following week we began vaccinating people at eighteen sites around the UK. We did not yet have the data on immune response – for that we would have to wait for the results of the blood tests taken at various intervals after vaccination. But we had enough data to give us high confidence in the vaccine's safety, so we were prepared to increase the pool of people in the trial.

We started our phase III trial in volunteers aged 18–55, which was the population we had tested in phase I. Now, as well as looking at safety and immune response, we wanted to test whether the vaccine actually worked. Does it stop you from getting ill with Covid-19, as opposed to just making your body generate the right kinds of antibodies and T cells?

At the same time, we started expanding the phase II safety and immunology trial to include older people. We vaccinated 160 people aged between 56 and 69, and then 240 people aged 70 and over. Again, we recorded all the safety data and the immune responses. Once we were confident that the vaccine was highly safe for these groups, we could add them to the phase III trial.

As had been the case for the phase I trial, we had no problems finding people willing to offer their arms to us. There were people who saw it as 'doing their bit', or a way to do something positive at such a difficult time. There were people who enjoyed the opportunity to get out of the house. Some people were still having to get close to vulnerable people as part of their work and wanted a chance to be protected and in turn protect others. Some had lost family members to

Covid-19. I think everyone recognised that they were part of something important.

This phase II/III trial that we started on 22 May used the first batch of vaccine from Advent: the batch that was puzzling us because it was giving us two different concentration readings when we used two different tests on it. At this point, our plan was still to give people just one dose. However, by mid-June the immunology data from the ten people we had given a second, booster dose, was starting to suggest that two doses would produce a better immune response. The people in this group had higher levels of antibodies, and we were finding similar effects in studies on mice, ferrets and monkeys. In early July we decided to change the trial design and give everyone two doses, at least four weeks apart. Everyone who had originally agreed to receive a single dose was contacted to see if they would be willing to have a second dose. That all took time, which is why we ended up with second doses being given at a variety of intervals. This turned out to be helpful, because it gave us information we would otherwise not have had about dosage intervals: when you want trial results as quickly as possible, you are unlikely to design in a twelve-week interval because of the delay it builds in. It also led to some confusion when we reported our efficacy data. We originally reported that the 90% efficacy result was in the group of people receiving a half-dose followed by a standard dose. But later analysis showed that giving two standard doses and increasing the interval between them from four to twelve weeks resulted in higher efficacy. It was this, not the low first dose, that was important. The decision to switch to giving everyone in the trial two doses at least four weeks apart also meant we would have to wait a lot longer before we could expect the results of our phase III trial.

For this part of the trial we also included a further type of testing: weekly swabbing. Volunteers in the UK were asked to swab their nose and throat every week, even if they had no symptoms. I don't know if you have ever had to do one of these yourself but I have, and it's not pleasant. I guess maybe you get used to it if you do it every week, but still, it shows commitment that we continue to receive, every week, thousands of these data points from our selfless volunteers. Volunteers post their tests to the UK national testing labs with a special barcode that marks them as trial samples, and the data generated is then sent to the trial team. In this way the trial was not only collecting the same data as the phase I trial – serious adverse events from health diaries, immune responses from blood samples, and symptomatic Covid from Covid tests – it was also building a picture on asymptomatic cases, which was important for understanding whether the vaccine could reduce virus transmission as well as disease.

Asymptomatic infection isn't a problem for the person who has it. They have a viral infection that doesn't make them ill. Unless they happen to do a test, they don't even know that they are infected. But, because people with asymptomatic infection don't usually know they are infected and so don't change their behaviour – but are still infectious – this type of infection is really important in spreading the virus to others, who might then become much more seriously ill. If the vaccine had no effect on asymptomatic infections the virus would continue to circulate in the population, even if everyone had been vaccinated. Since no vaccine is 100% effective, the virus would then continue to make a small proportion of people ill.* Also, if the

* Why are no vaccines 100% effective? Because vaccines rely on the immune system and some people's immune systems don't work very well – for example, people with untreated HIV or people who have been having chemotherapy, and older people. As mentioned earlier, as people age, their immune system becomes less effective.

virus continues to circulate, it will mutate, and could do so in a way that makes it more transmissible or more dangerous or more able to evade a vaccine.

The swabs that our volunteers sent in every week enabled us to build up a picture of how many people had infections, and for how long, and with what amount of virus in their nose and throat, despite not having any symptoms.

We were vaccinating more people and the dosing schedule had become more complicated. Space at our little clinic was becoming an issue. We had already expanded the trial to other centres in the UK, but that still left us with a problem in Oxford. One day in June, our buildings and facilities manager, Oto Velicka, emailed me. 'Cath, is it OK if we hook the new clinic up to the CBF's power supply?'

'What new clinic?' I messaged back. Oto suggested I look out the window. It was like something from a sci-fi movie. Over the course of an afternoon, truck after truck crawled down the gravel track, performed some graceful manoeuvring, and unfurled like a strange white flower. Within forty-eight hours a fully functional vaccination and testing clinic had emerged in the car park.

However, even with our shiny new pop-up clinic, there was no way that the team in Oxford could manage the numbers needed for a phase III trial. Also, the case rate in Oxford was very low during the summer of 2020 when these trials were starting, whereas other parts of the country were still suffering from significant levels of infection. Remember, we need some people in the trials to be exposed to the virus, to know whether the vaccine works. Sarah was receiving lots of emails telling her to go to Birmingham, or Leicester, where infection rates were high. Unfortunately that didn't really help. By the time we had set up, recruited volunteers, vaccinated them twice and waited for the vaccine to induce an immune response, case rates would have been brought down again. We did though have the trial set up in eighteen sites around the UK, and we were prioritising healthcare workers who were more likely to be exposed to the virus in the course of their normal lives.

There had also always been plans to test the vaccine in other countries. The whole world was going to need it, not just the few countries that could afford to pay the most. Immune response can be influenced by ethnicity, and also by the different microorganisms that people encounter in their everyday lives. In some countries, regulators will not allow a vaccine to be used unless there has been some testing in those countries. But also, with trials spread around the world there was a higher chance that at least one of them would be somewhere with high transmission: bad for the local populations, but good for the trial. In June, once we had shown in UK volunteers that the vaccine was behaving as expected in terms of reactions and immune response, we expanded the trials to Brazil and South Africa. A lot of countries had come forward to say they wanted to run trials, but we still had only limited quantities of vaccine. We chose these countries because we already had good links with highly professional and experienced colleagues there and were confident the trials would run smoothly.

—

I hit peak stress at the start of May. I was trying to juggle a lot of problems that felt out of my control and like many people I had had no physical contact with another adult in months. Until this year I had completely underestimated the power of hugs. I know my friends were worried about me because they offered food deliveries and care packages, and bunches of flowers started showing up unexplained on my porch.

However, over the summer, life got back to a strange version of normal. At the CBF we had our glum days, because there was a lot of work to catch up on, and we were still socially distancing and missing the normal social interactions that glue a team together. But there was a feeling that things were gradually getting better. People came out of their houses. There were kids larking about in Port Meadow again, and people boating on the river, while my daughter Ellie and I were swimming in it. Pubs were allowed to open, and we could sit in small groups in the garden at the Perch or the Isis with a cold beer.

I even managed three stolen camping weekends away, though in each case this did involve driving around to find a 4G signal so I could do Zoom calls from the car whilst Ellie was tasked with keeping an eye out for traffic wardens. I love living in a city and can't be doing without bars and a café to make me a good brunch, but I also love the feeling of sleeping under canvas and don't mind if it's cold and wet so long as there is a campfire and some rum. It was also lovely to spend some time just hanging out with my daughter, swimming and eating ice cream. The places we stayed had made huge efforts to be Covid-secure, with fewer pitches and lots of cleaning in the shower blocks. As well as the camping trips, I took advantage of Eat Out to Help Out, saw my mum and dad for the first time in months, and ran a rule-of-six socially distant mock-Glastonbury Festival across multiple gardens. Still no hugs though.

On 18 July 2020, fifty-six days after the last of our phase I volunteers had been vaccinated, the trials team had an initial set of data on their immune responses. Waiting for the immunology results had felt like waiting for my A-level results: I knew that I had worked hard and I was pretty sure it had gone OK, but I just needed to see the piece of paper. Until then there was always that nagging doubt: what if I read the question wrong? Similarly, with the vaccine, we knew it must be highly safe. Hundreds of people had received it and if there had been any serious issues, the trial would have been stopped. We had got strong immune responses when we had tested the vaccine on mice in the lab. We knew that the MERS vaccine it was modelled on had been very safe, too, and had induced a good immune response. So we had our expectations. But we just needed to see the piece of paper. And the moment Andy showed us the graphs that demonstrated the right kind of immune responses, we knew we had passed (and with a good grade!). This was hard evidence that the vaccine we had made actually had a chance of working. It was difficult not to tell my friends, who were asking for news every day. Being tight-lipped is not in my nature (though I do manage it when required, which has been quite a lot in the last year). But within a few days the team had published a paper with their findings and we were free to talk about it.

The paper set out the results of our first phase I/II study and it was very good news. Its opening line was, in its own understated, scientist way, thrilling: 'The pandemic of severe acute respiratory syndrome coronavirus 2 (SARS-CoV-2) might be curtailed by vaccination.' As one journalist wrote in the *New Statesman* the next day: 'it is a cautious good news story, with the potential to become such a good news story it almost doesn't bear thinking about'.[2]

There had been no serious adverse events related to the use of ChAdOx1 nCoV-19. The vaccine was very safe to use. Just as important, the lab tests showed that the blood of the vaccinated volunteers contained high levels of the antibodies and T cells that should help to fight off Covid-19. And in the small group that had had a booster, the levels of antibodies went up further after the second dose. This was a strong suggestion that the vaccine would work to protect people against the disease.

But the number of volunteers who had got sick with Covid-19 by this time was too small for us to be able to tell whether that would truly be the case. We also had no data yet on immune responses in older people. So really this was less like getting A-level results and more like doing really well in your mocks. It boded well, but the real test was still to come.

When I look back to the day of that first vaccination, 23 April 2020, it seems such a long time ago. We hadn't yet had the huge challenge of the 2020–21 lockdown cycles of hope and despair. On the radio I felt able to send out a message of hope that the trial was starting and the vaccine was coming. That evening, Tess, Sandy and I got together for fizzy wine by Zoom.

As I write, after an unbelievably hard year when many of us lost loved ones before their time, we are back to hope again. The trials have proved that the vaccine is very safe and highly effective, and it has started to be used to protect our families and friends. If you volunteered for this trial, thank you. This is how scientific progress is made. Thank you too to volunteers in previous trials, for MERS, or malaria. And thank you to volunteers in trials for other Covid vaccines. Every trial builds the knowledge base for all future vaccines. Every trial contributes to scientific progress and potentially allows us to move further and faster the next time.

CHAPTER 9

The Prince and the Protestors: Vaccine Acceptance and Hesitancy

Conservative estimate of lives saved due to vaccination, 2000–2019: 37 million[1]

We don't often come into direct contact with anti-vaxxers, although a few did make an appearance on one particularly memorable day in 2020. In June, Andy told me that we'd be meeting a very important visitor the following week. I asked Andy if it was a politician. I was of course extremely grateful for the funding we had received from the UK government, but a little concerned that we would now become a 'must-visit' destination for a long line of politicians wanting to drop by for a photo opportunity. 'You have to promise not to tell anyone – not at work and not at home,' said Andy. 'It's Prince William.'

The visit was exceptionally carefully planned and at 8.30 on the day itself, Andy and I had a briefing call with the university security team to run through the arrangements one last time. Security had received news of a group planning to stage a protest against our vaccine trials that morning, which was a slight concern. The protestors didn't know about the royal visit of course, and they were planning to target my building, rather than Andy's or Cath's, where it was actually due to take place. It seemed they were being led by one person who had staged protests about a variety of things in the past: fracking, the anti-coronavirus lockdown and Covid-19 vaccines. Every so often he would announce his intention to stage a protest on his Facebook page and invite others to join him. His previous one had been at another of our clinical trial sites, and had been a slight irritation rather than a security risk. He was apparently asking why we were vaccinating children, even though we weren't.

Security was not expecting a very large crowd for the protest, and my office looks out over the front of the building, so I suggested that I could just check for any protestors by looking out of the window before setting off to meet the prince. If necessary, we could leave by another exit and walk the long way round. The security team was not entirely happy with that plan, but we agreed to speak if there seemed to be a problem.

When it was time to leave, Tess and I had a good look out of the window. It was a glorious June day – lovely weather for a protest. There was no one directly outside, but a bit further along, a bunch of men in dark jackets were standing around a car looking shifty. Tess pointed out that that was actually the university security team. So we left by the front entrance and headed over to the clinical centre. First stop was the PPE station, where we sanitised our hands and put on surgical masks. It was the first time I had worn a mask. In my building at that point we were relying on social distancing achieved via very low staff numbers, but at the clinic and CBF they had needed to implement strict PPE protocols early on, because of all the trial volunteers coming in to that part of the campus. It took me longer than I expected to put it on and adjust it to fit well around my nose. (A few months later, it had become second nature to put on a mask before entering my building, and every time I left my office.)

Andy and I were going to be having a chat with the prince that would be filmed in a fairly unglamorous meeting room on the ground floor.* After introductions but no handshakes, the three of us sat down. It was a slightly odd experience, the three of us sitting in a row, two metres apart, being filmed, but it probably wasn't the oddest thing that happened that year, and Prince William was well informed and seemed genuinely interested and supportive. Afterwards, I stayed in the meeting room and Andy led the way to the CBF. Cath told me later that six minutes had been assigned for a tour of her clean room where the vaccine is made. The prince was a good sport about doing introductions and then getting himself signed in, into clean-room clothes (overshoes, gown, gloves), through the airlock, into the clean room and back out again in the allotted time – all while retaining everybody's name. Then it was on to the mobile clinic to drop in on some people who were being vaccinated, upstairs to the lab to meet Tess and members of the lab team, and then to another odd-looking event that I was able to glimpse through the slats of the meeting-room blinds. Prince William was standing in the courtyard in a marked square on the grass

* There were vertical strip blinds at the windows which were going to form our backdrop but they were old and damaged and not hanging smoothly. I was left with the job of sorting them out while everyone else went outside again to greet the prince. I sat down on the floor to assess the worst section and became involved in the task, diagnosing how many strips would need to be unclipped and reclipped to their beads and then doing the best job I could of repositioning everything. It wasn't perfect – there wasn't much I could do about broken clips or squashed beads – but I managed to get the strips arranged evenly and was just leaning back to examine my handiwork when I heard voices in the corridor. I jumped to my feet just as the prince and the press entered the room with Andy. I was thankful that I hadn't still been sitting on the floor fiddling with the blinds, and also for the mask that covered my non-composed expression.

facing and chatting to four members of our team, each also in their marked square. Other people were watching from the edge of the courtyard, all smiling broadly. The whole thing was a great morale boost for the team.

I later found out that a few protestors had turned up at my building and shouted a bit, but they had arrived, shouted and left all during the time I was at the clinic. If they watched the news that evening, they would have seen that only a few hundred yards away from their anti-vaccine protest, Prince William's supportive visit to the vaccine centre was taking place. They must have been kicking themselves.

—

I don't understand anti-vaxxers. Why would anyone be ideologically opposed to a safe and cost-effective public health measure that saves millions of lives and stops people from having to live with the long-term disabilities that can be caused by diseases such as polio and smallpox – and, it seems, Covid-19? But there will always be people with strange beliefs who stick to them come what may. Anti-vaxxers, who are actually a pretty small, if noisy, group, only become a problem if they try to interfere with other people's acceptance of vaccines.

Vaccine hesitancy, however, is a different matter. It is natural that people want to understand the risks and benefits of vaccines, and important that as scientists we engage with their concerns.

It has been very common over the last year or so to read or hear things like 'we don't know what's in it', or 'is it going to change my DNA?', or 'it feels like it's been rushed and that makes me uncomfortable', or 'I'll take my chances with the virus, thanks'. When I catch myself bristling at yet another question or comment that implies I and my colleagues are up to no good in some way, or conducting poor science, I try to remember that there is a history that – even if people do not remember the details – informs these concerns.

When Maurice Hilleman developed his mumps vaccine in record-breaking time for example, in the 1960s, he did some things that we would consider at best ethically questionable today.[2] Mumps is generally a fairly mild disease in otherwise healthy children. But it can sometimes cause meningitis resulting in seizures, paralysis and deafness, or infect the pancreas and cause diabetes. It can also cause birth defects if pregnant women are infected. Hilleman conducted the first clinical trials of his vaccine on sixteen children at a home for children with learning disabilities. He went on to test it at another children's home, and then on hundreds of children in nursery schools and kindergartens in Philadelphia where he was based. Today, we would always test vaccines in animals before testing them in humans, and in adults before testing them in children. We also ensure we have the informed consent of everyone participating in a trial, or if they are children, the consent of their parents or guardians – which was not obtained for the children in Hilleman's first trial.

This is not the only example of people being used in medical research without their knowledge or consent. The infamous Tuskegee experiment in the United States recruited 600 African American men into a study and told them they would receive free healthcare. The study lasted forty years, from 1932 to 1972. The real purpose of the experiment was to observe the course of untreated syphilis infection in the men, of whom two-thirds were selected for the study because they had latent (meaning currently asymptomatic) syphilis infections. The men were not told of their diagnosis, and were never treated with antibiotics, which were widely available by 1947. During the study, 128 of the men died either from syphilis or related complications. Forty of their wives caught syphilis and nineteen children were born with birth defects.[3] The legacy of this appalling experiment is still felt today, with many African Americans and people from other ethnic minorities reluctant to take part in medical research for fear of being lied to and experimented on.

The Tuskegee experiment was not a vaccine study. But it was also not a one-off. It was perhaps the most egregious example of many exploitative experiments that were carried out on children, prisoners, senile patients and other vulnerable groups of people in the United States in the 1940s, 1950s and 1960s, in pursuit of the development of both drugs and vaccines.[4] The exposure of these studies, and in particular of the Tuskegee experiment, led to legislation to prevent such things happening again. Any type of medical research on people, including vaccine studies, now requires approval from independent ethics committees, and informed consent from and appropriate communication with all participants.

Other aspects of vaccine development have also caused problems in the past. In 1796 Edward Jenner, considered the father of immunology and the person who has saved more lives through his work than any other human, famously vaccinated James Phipps, the 8-year-old son of his gardener, with pus taken from a cowpox blister on the hand of a milkmaid, Sarah Nelmes.[5] For decades afterwards, vaccination did not require the production of vials of vaccine as we now know them, but was 'arm to arm', transferring material from a blister on the skin of someone who had been vaccinated to the next recipient. Unfortunately, sometimes more than cowpox was transferred. In the second half of the 19th century, a group of children in Italy developed syphilis, and in Germany there was a large outbreak of hepatitis.[6] Even when vaccines started to be produced and put into vials in the 1940s, sometimes things went wrong. In 1942, human serum being used to stabilise a yellow fever vaccine for American servicemen turned out to be contaminated with hepatitis B virus. Around 50,000 of those given the vaccine were hospitalised with hepatitis, and at least one hundred died.[7] In 1955, US-based Cutter Laboratories, producing an 'inactivated' polio vaccine, failed to inactivate a batch fully, resulting in polio infections in those vaccinated; 192 people were paralysed as a result, many of them children, and ten people died.[8]

These events still linger in our folk memory and lead to a general sense of unease around vaccines. However, they all happened more than forty years ago, and scientists and regulators working on vaccines today have learned from that history. Measures have been put in place to ensure none of those events could happen today.

The possibility of contamination with a live pathogen, for example – either the pathogen being vaccinated against or another pathogen – is now something that is controlled for at every step of the manufacturing process. When using platform technologies such as ChAdOx1 or DNA or mRNA vaccines, rather than starting from the pathogen itself, there is no risk of failing to inactivate the pathogen being vaccinated against (as with the polio vaccine in 1955) or to weaken it sufficiently. The platform technologies increasingly in use in the 2020s are either not live at all, or live but replication-deficient, meaning they can therefore never spread through the body and cause infection and disease.

As Cath describes in Chapters 3 and 5, the manufacturing process is tightly controlled and all raw materials are tested before the vaccine is made, with many more tests to check that no contamination with anything else (as with the yellow-fever vaccine in 1942) has occurred at every stage. All of the information on exactly how that has been done must be provided to the regulatory authorities to receive permission to start a clinical trial, let alone apply for a vaccine to be licensed for use.

There are also strict ethical controls. Alongside the regulatory application to run a trial or license a vaccine, there is an ethical application. Here, those conducting the study must provide all the information they wish to pass on to volunteers, including any material used to recruit volunteers, an information sheet, and a consent form to sign. The volunteer information sheet will be scrutinised for clarity as well as accuracy, aiming for a reading age of 12 to ensure that all volunteers will fully understand the information they are being given. The consent form will be checked to ensure that it is fair and reasonable.

For studies involving children, there are separate information sheets for parents and children, and sometimes different sheets for children of different ages. Children may be asked to 'assent', meaning they have a say having been given some information they can understand, but parents can still refuse permission, giving them the final say. For more complex studies such as malaria vaccination challenge studies, in which volunteers are deliberately infected with malaria to find out if the vaccine protects them, there is a questionnaire for the volunteers to take to find out whether they have understood the information they have been given. It is a key requirement of those studies for the volunteers to come to the vaccination clinic twice every day after they have been infected, so that they can be closely monitored and then treated with anti-malarial drugs when necessary. Failing to understand that could be dangerous, and so researchers will only accept someone onto the trial if they are confident they have understood the information.

One ill-informed comment I have read this year is 'well, let's hope we don't have another thalidomide!' Thalidomide is a drug used to treat a range of conditions including cancer and complications of leprosy. A side effect is sleepiness and when it was first introduced it was promoted to prevent morning sickness in pregnancy. In the late 1950s and early 1960s, it took five years and the birth of 10,000 affected babies to establish that thalidomide, though safe for most people, caused birth defects if taken in pregnancy. Many changes were made to the way drugs were tested, approved and marketed after the thalidomide scandal. For example, in the UK, drugs intended for human use could no longer be approved purely on the basis of animal testing but had to be tested in humans too. Drugs marketed to pregnant women had to provide evidence they were safe for use in pregnancy. More controls were put in place around over-the-counter access to medicines. And the Yellow Card Scheme was put in place for doctors and later anyone to share side effects of medications they prescribed.[9]

Today, new vaccines are never tested in pregnant women until there is a great deal of information about their safe use in non-pregnant women (and in men). The next step is to conduct developmental and reproductive toxicology (DART) studies in animals, vaccinating either before conception or soon after conception, tracking the progress of the pregnancies, and waiting for the offspring to be born and then studied. In early-stage clinical trials, women of childbearing age are required to use effective contraception for the duration of the study, and are tested before each vaccination to make sure that they are not pregnant. After vaccine safety has been demonstrated in the general population and the DART studies have been completed, a separate study in pregnant women will be conducted before the vaccine is recommended for use in pregnant women generally. Of course there are many vaccines that are safe for use, and recommended for use, in pregnant women, but this is something that is carefully assessed rather than assumed.

Nonetheless, and even with everything that has been done to address and avoid the mistakes of the past, it remains the case that nothing in life is risk-free. And so, when deciding whether to receive a vaccine, as with everything else in life, we have to decide whether the benefits outweigh the risks.

Unfortunately, it is also the case that as human beings we are just not good at understanding different levels of risk. We tend either to think things are *very worrying*, or to believe that they are *absolutely fine*, when in fact nothing we do, or fail to do, is risk-free and most things are somewhere in between.

For example, many people would not go out picking wild mushrooms for dinner in case they picked a poisonous one (*very worrying*). A 'death angel' is an innocuous-looking white mushroom that can cause liver failure and death. But many of the same people would give no thought at all to eating a wild mushroom risotto at a restaurant (*absolutely fine*). In fact, though the restaurant is less risky, it is not risk-free. Did the restaurant test each batch of mushrooms for toxicity? Are they from the same supplier every week? What happens when the expert forager goes on holiday and leaves someone else to do the picking for a week? Both picking our own and trusting someone else are somewhat risky.

We face the non-zero risk of being injured or killed every time we leave our home, but because the benefits of leaving the house are so great, we tend not to spend much time thinking about or trying to calculate these risks (*absolutely fine)*. Plus, there are also risks to staying in the house. A plane might crash into the bedroom window, or, more likely, a candle might start a house fire.

We also tend to overestimate risks of things that are rare but dramatic, like plane crashes, and underestimate the risks of things that are so common that they tend not to feature in the news, like car crashes. Whether you measure it by distance travelled or hours spent on the journey, travel is much safer by air (or bus and rail) than in a car.[10]

A logical (but exhausting) approach would be to assess the risks and benefits of every one of our actions or inactions, and then reduce the worst risks as much as possible, ideally without losing the benefits. To conduct a thorough risk assessment, we need to think about both the *severity* and the *likelihood* of an event. A freak occurrence like a plane crashing through the bedroom window is high on severity but low on likelihood. Since it is very rare and also there is nothing we can do to make the already tiny risk smaller, we shouldn't spend any time thinking about it, and certainly shouldn't avoid sleeping in the bedroom in order to mitigate the risk. Getting stung by a wasp in the garden now and again is highly likely, but for most of us not very severe. Again, we probably shouldn't spend time worrying about it, and should just enjoy time in the garden. (If we know we are allergic to wasps, though, we should make sure we have an EpiPen to hand.) A car crash is going to be somewhat severe and somewhat likely. In this case, it is worth being aware of, and trying to reduce the risk, for example by keeping the vehicle properly maintained, wearing a seat belt, not drinking and driving or using your phone, and obeying the rules of the road. This will reduce the risk but not remove it completely: because of other road users, accidents can still happen to highly competent drivers in well-functioning cars. Most of us will nonetheless sometimes opt to put ourselves and our children in cars because of the benefits they bring.

When doing our own, personal, risk-assessment on a vaccine, we should bear in mind that a great deal of risk-reduction has already taken place. Every stage of the process is tightly regulated. The raw materials we use are from suppliers working to specific quality standards but are tested again once received at the manufacturing facility. The facility itself must satisfy many requirements and is periodically inspected by regulators, who also want to know about all of the manufacturing procedures that are used and how the staff have been trained. Trials take place first in animals then in small groups of healthy adults and from there in increasing numbers of adults, to ensure the vaccines are highly safe before they can be licensed.

We should also bear in mind that not all vaccines are the same. One common misconception around vaccines is that they all contain lots of things that we would not want injected into us. In the past I have found it frustrating, when trying to address these concerns, that some people lump all vaccines together. For example, my brother-in-law contacted me to say his friend was hesitant to have the flu vaccine because, she had said, 'of course it contains mercury'. This is not true. Many years ago, when flu vaccines were often put into multi-dose vials, they sometimes contained thiomersal, a mercury-based preservative. If bacteria were accidentally introduced when the vial was first used, the thiomersal would kill them and prevent them from being injected into anyone receiving a later dose from the same vial. However, thiomersal has not been used in the UK for many years. None of the vaccines in routine use in the UK, or Europe or the US, contain thiomersal. Even when they did, the amount was tiny, and the mercury was in a form that was quickly excreted.[11]

Thanks to the media interest in different kinds of Covid vaccines, I hope it is perhaps now clearer that there are many approaches to making vaccines, and as a result different vaccines will have different ingredients. You can easily find out what ingredients are in any vaccine you are concerned about, because every vaccine is accompanied by a 'Summary of Product Characteristics' sheet that sets out, among other things, exactly what is in it. You can find them by searching online, and for our Oxford vaccine you can turn to the back of this book and find an annotated ingredients list.*

Many vaccine ingredients might look unfamiliar but are in fact substances that are found naturally in the body. All are present in extremely small quantities (a few thousandths of a gram or less) and other than for people with severe allergies there is no evidence that any of them cause any harm in these amounts. There is a very small amount of alcohol in the Oxford AstraZeneca vaccine, at around the same level that is found in some natural food. (The British Islamic Medical Association statement on the vaccine says: 'This is "not enough to cause any noticeable effects" and has been described as negligible by Muslim scholars.'[12]) When Covid vaccines began to become available, there were messages circulating on social media saying that they contained substances derived from pork or beef, and that they were therefore not suitable for Muslims or Hindus. This is not true, just as the claim that flu vaccines contained mercury was not true. But messages from friends are generally trusted and passed on, and the correct information is not often sought out. It took further social media campaigns to convince people that there was no basis for these claims.

It can be hard to shift someone's long-held view even if it is wrong. I sent my brother-in-law the ingredients list for the flu vaccine his friend was concerned about, for him to pass on.

---

* Appendix C sets out exactly what is in the Oxford AstraZeneca vaccine. You can also find out what is in other vaccines, together with explanations of what each ingredient is and why it is in the vaccine, on the Vaccine Knowledge Project website.

But I wonder whether that will have changed his friend's mind. Perhaps she found it hard to accept that she had been making an incorrect assumption for years. Perhaps she took offence at being told she was wrong. She felt she 'knew' there was mercury in flu vaccines (there isn't) and that this was harmful because it was toxic and built up over time (the form that used to be used in some vaccines doesn't do this).

However, aside from mistaken worries based on incorrect information, there are three kinds of risks we do need to consider and weigh up when it comes to vaccines.

First, not all vaccines are safe for everyone. Some vaccines are not advised for specific groups of people. For example, some vaccines use weakened versions of the pathogen itself, or of a closely related organism. They include the BCG vaccine used to prevent tuberculosis, and the MMR vaccine against measles, mumps and rubella. Vaccines made this way are known as 'live attenuated' vaccines, but now that more types of vaccine are being brought into use it would be more helpful to describe them as live attenuated *replication-competent* vaccines. These vaccines spread within the body after vaccination. They do not cause disease in a healthy person, and the immune response that the body makes to quickly control the infection is what results in the formation of immune memory that can then protect against the pathogen itself.

However, in people with severely compromised immune systems, when the vaccine starts to spread through the body the immune system is unable to control it, and the infection may become serious, or even fatal. These vaccines are therefore not used in certain groups of people (for example people with HIV, or people undergoing chemotherapy). By contrast ChAdOx1-vectored vaccines are 'live' but are replication-deficient. The adenoviral vector infects cells in the body soon after vaccination and instructs the cells to produce the SARS-CoV-2 spike protein, and then it has done its job. It is not able to make any more copies of itself, and cannot spread through the body even if the person being vaccinated has no functioning immune system. Even severely immunosuppressed people can take the ChAdOx1-vectored vaccine with no risk of uncontrolled infection.

mRNA vaccines like the Pfizer-BioNTech and Moderna vaccines are not live. They are briefly present in some cells in the body, but cannot spread.* Some people have expressed concern that the vaccine could permanently change their DNA but this is not true. The mRNA used in the vaccine never enters the cell's nucleus which is where your DNA is kept, and the cell breaks it down and gets rid of it once it has used it.

Another group of people for whom not all vaccines are safe are those with some types of severe allergy. Some influenza vaccines, for example, are produced in chicken eggs, and therefore contain tiny amounts of egg protein. This can cause an allergic or even anaphylactic reaction in people who are severely allergic to eggs. People with allergies to eggs or antibiotics should always ask about the presence of trace quantities of these in any vaccine they take (neither is present in the Oxford AstraZeneca vaccine). There is usually an alternative that can be offered.

On the first day of the roll-out of the Pfizer-BioNTech vaccine, two healthcare workers who had been vaccinated suffered allergic reactions. Both of them had known allergies and carried EpiPens.† This is the kind of thing that can happen when a vaccine that has only been used in clinical trials in which people with severe allergies are excluded is then used in a much

---

* There is more about how mRNA vaccines work in Appendix A at the back of the book.

† The most likely component of the vaccine that might be causing these responses was thought to be polyethylene glycol, which is also present in the Moderna vaccine.

wider population. As a result the MHRA said that people with known severe allergies should not receive the Pfizer vaccine, and all others should be observed for fifteen minutes after vaccination.

And, as already discussed, vaccines should not be given to pregnant women until appropriate trials have been conducted. An exception would be when there is a high risk of infection. In this case, there should be a discussion about the risks and benefits for that individual.

There is lower take-up of Covid vaccines amongst people from some ethnic minority backgrounds. The reasons are complex and it is important not to be simplistic in assigning reasons for vaccine hesitancy to large and diverse groups of people. However, in a review of vaccine acceptance in the UK, one of the concerns raised was that vaccines might not have been tested in people from different ethnic minorities, leading to fears that vaccines had been developed for one enthnic group and might not be safe or effective for others. It is important, for that reason, that clinical trials of vaccines do include people from a range of different ethnic backgrounds. This was addressed in the SARS-CoV-2 vaccine trials that took place in 2020 and 2021.[13]

The second kind of risk to consider is that vaccines have side effects. Most side effects, or adverse events, are mild, occur very soon after vaccination, and are over within a few days. These – sore arm, a fever, fatigue, headache – will be picked up during clinical trials and fully documented. More severe adverse events also tend to occur soon after vaccination if at all, and also will be picked up in a trial. At that point, the trial may be paused while the event is investigated. Following thorough investigation, the trial may be stopped completely, it may be started but only for certain groups of people, or it may continue. But there is also a possibility of some very rare side effects, which, because they are so rare, do not occur during the trial, and are only identified after millions of people have been vaccinated. This was true of the H1N1 influenza vaccine produced for the swine flu pandemic of 2009–10.[14] Studies eventually showed that the vaccine was likely to have been responsible for three cases of narcolepsy (a chronic sleep disorder where people fall asleep at inappropriate times) for every 100,000 children and young people vaccinated in northern Europe. The effect wasn't found in the rest of the world, because narcolepsy was induced only in those with a particular gene that is more common in northern Europe. No phase III trial would ever have been large enough to detect that effect. Once a vaccine is licensed and used widely, there is a safety reporting scheme, which does then pick up very rare problems and take the necessary action.

This is what happened in March and April 2021 when reports started emerging of a very small number of people developing a rare type of blood clot soon after being vaccinated with Covid-19 vaccines. It was possible that these blood clots had been caused by the vaccine. But it could not immediately be established for certain: because these events were so rare, and could occur naturally even if vaccination hadn't happened, it was very difficult to understand what was going on, and why.* What was clear was that the risk, if there was one, was

* I was of course distressed that our vaccine might be causing harm to even a tiny number of people, and as a scientist I set about trying to find a plausible explanation or biological mechanism for these rare events. If we assumed there was a causal link between vaccination and blood clots (although this was not proven), was that related to some fault in vaccine manufacturing? Or to the ChAdOx1 vaccine vector itself? Or to the spike protein – which might mean all of the vaccines would be affected to some extent? What did we know about adverse events in other Covid vaccines? Unfortunately, systematic reporting of adverse events was not as thorough in all countries as it was in the UK and so there were gaps in the information available. I spent much of the Easter weekend in my office reading, thinking and trying to design experiments to aid our understanding. Ultimately, any explanation would need to be verified independently and Oxford scientists would not be able to do this work alone, but if I could come up with a plausible hypothesis to test, it might help.

vanishingly small. Indeed, it is known that a confirmed risk of a fatal blood clot from a common drug that people took for years at a time, the contraceptive pill, was much higher. The risk of a fatal blood clot after being infected with Covid-19 was much, much higher.

Regulators and policymakers in the UK and elsewhere took the new information about these very rare events and carefully considered the implications. They had to take into account not only the potential risk from the vaccine, but also its known benefits. (Or, to put it another way: the risks of taking the vaccine, and the risks of depriving people of it.) These varied depending on circumstances. For example: how much virus was circulating (the more virus circulating, the higher the risk of catching Covid and so the greater the benefit of a vaccine); what age groups were being considered (the older the person, the higher the risks from Covid and so the greater the benefit of a vaccine); and were there alternative vaccines available to use (is the decision this vaccine or no vaccine, or is it this vaccine or another vaccine)? Both the UK and the EU regulators said that the risk–benefits were hugely in favour of continuing to use the vaccine. SARS-CoV-2 had dealt the world a terrible blow, with millions of deaths so far and continuing misery for those suffering from long Covid that we were only just starting to understand, and vaccines were our way out. Lockdowns could reduce infections but came with their own costs to people's health and well-being. Case numbers were still rising in Europe, Latin America and South Asia, Covid was still killing one in every 150 people infected, and the vaccines that protected against it were doing vastly more good than harm: the day after the MHRA announced that nineteen people had died from rare blood clots in the UK that might have been due to vaccination, the latest analysis showed that vaccination in the UK was saving hundreds of lives every day. The more difficult analysis was whether the risk–benefits remained clear for all parts of the population and in every country. In early April, based on a calculation that there might be a 1 in 250,000 chance of developing a rare blood clot after vaccination, and a 1 in 1 million chance of dying from it – the same risk as driving 250 miles in a car – the UK decided to offer an alternative vaccine to adults under 30. The reasoning was not that the risk to under 30s was higher. It was that the benefit of the vaccine to this age group, who were much less likely than older people to die from Covid-19, was lower, particularly now that there were lower case rates, and alternative vaccines available.* Trials of the vaccine on children were also paused, not because of any specific safety concerns, but on the basis that this would allow time for further information about safety in adults to be assessed. And advice was provided about symptoms of these very rare blood clots, so that anyone experiencing symptoms could alert their doctor and receive treatment.

As circumstances change, and as we learn more about these rare blood-clot events, it may be that further adjustments need to be made. I cycle to work every day, on a route that I consider to be safe. However, for the past few months this route has taken me past a white-painted bike marking the spot where a young woman cyclist was killed in a collision with a bin lorry. Even knowing about this distressing event, I continue to cycle to work, and my children also often cycle that route. For me, the benefits to my health and my mood and to my ability to get where I want outweigh the very small risk of death. As

* This was based on a very conservative assessment of the benefits of the vaccine. It did not for example take into account the benefit of avoiding long Covid. Nor did it take into account the benefit of not transmitting the virus to others.

circumstances change, though, so does my risk assessment. When it is icy, I walk or drive instead. When there is snow, I walk. Rare and dramatic events like blood clots or plane crashes can make people very worried. But, as discussed above, nothing in life can ever be completely risk-free, not even sitting quietly in your house doing nothing.

The final theoretical safety concern is that some vaccines have resulted in a severe form of the disease when the vaccinated person is later exposed to the pathogen. As discussed in Chapter 8, now that scientists know about this concern and understand what kind of immune response is involved, we are able to conduct preclinical (animal) trials to assess this risk. We cannot, however, be absolutely certain that we have ruled it out in humans until some of the volunteers who have been vaccinated have been exposed to the virus as part of their normal lives. Thanks to the very large phase III trials that we have carried out, that too has now been thoroughly studied. All of those people on our trials who received the vaccine but still became infected (which happened on all the trials because no vaccine is 100% effective) experienced only mild disease.

Taking a vaccine is not risk-free – because nothing is. But not taking a vaccine is also not risk-free. You risk getting the disease yourself, and also passing it on to others.

Many people, faced with all the difficulties weighing up these risks, prefer to trust their healthcare professionals. That is a very reasonable approach. The regulatory authorities, when deciding whether to approve a vaccine for general use – or, in a pandemic, for initial emergency use – will have done their own very thorough risk–benefit analysis, and provided comprehensive guidance to healthcare professionals including about whether there are particular risks or benefits for particular groups of people such as the elderly or the young, pregnant women, or people with severe allergies. The experts will continue to update their analysis and guidance as new information becomes available, for example in relation to rare side effects.

Others decide to find out more for themselves. That is also a very reasonable approach. But if you are going to do this, think about where your information is coming from. If you volunteer to take part in a vaccine clinical trial, all of the information that is given to you will have been carefully checked and approved by the regulatory authority and an ethics committee who are qualified to check both the information and the way it is presented. If you search for a Summary of Product Characteristics sheet online, the document you will find, describing a given vaccine, its risks and benefits, and who should not take it, will be long and tedious but it will have been approved by the regulators, with every statement reviewed and verified. It will never, of course, simply say 'this vaccine is completely safe'. Because – to labour the point – absolutely nothing is. Next time you take paracetamol, have a good look at the information sheet provided in every pack. There is a long list of risks documented there.* But most of us rarely even unfold the piece of paper.

If you rely on friends and family or social media instead for your information, there is no telling what you will find. When the information comes from someone we know, and who we think has our best interests at heart, or whose other views we respect, we might be tempted to accept it without questioning it – even to send it on to all of *our* friends and family. But it

* For example, twenty 500 mg tablets can cause liver damage. Ten tablets can cause liver damage to someone who regularly consumes more alcohol than recommended. Adult dosage is one to two tablets every four to six hours, so there is not a huge difference between safe and unsafe dosing.

is always worth asking where their information came from, and then making your own assessment.

There are virtually no checks on what people can say on social media. Some people will post misinformation in good faith, genuinely believing they are right. We have all probably received this kind of thing from well-meaning friends and family members. Others, often celebrities or influencers with big followings, do so because they are paid to, or are after clicks. Still others, sometimes hostile states or their proxies, maliciously post false information to deliberately confuse and mislead people. One group of people who may be acting in good faith but are wrong about vaccines are those who promote 'natural immunity'. Not all 'natural therapies' are dangerous or without value. Exercise and a better diet will often be of benefit to people presenting with, say, a kidney problem or a bad knee. Breathing exercises and talking therapies can be good for mental health. But 'natural health' does not provide all the solutions to treat or prevent illness. Aromatherapy massages may be offered to cancer patients undergoing chemotherapy, and they can provide pain relief by triggering pleasant sense memories and promoting relaxation. But it is the chemotherapy that is providing the cure, not the massage.

Vaccines may be dismissed by proponents of natural health as being 'non-natural'. Unfortunately, the only way to 'naturally' gain immunity to an infectious disease is to become infected with that specific disease, and that is not necessarily a safe thing to do. Maintaining our general health including a functioning immune system is a worthy aim, but we cannot develop the specific immunity needed to protect us from an individual pathogen without either being infected with it, or being vaccinated against it.

Another message doing the rounds on social media in 2020 and 2021 was that Covid vaccines would cause infertility. This was sometimes backed up by a quasi-scientific-sounding claim, that the Covid-19 spike protein was similar to a protein found on the placenta, and that antibodies against the spike protein would therefore also attack the placenta, resulting in infertility. There were no grounds for this claim. To think through its implications, if it were true, any woman infected with SARS-CoV-2 would develop antibodies to the spike protein and become infertile. This has not happened. There are also four other coronaviruses that infect humans, circulating mainly in the winter months, each with their own spike protein. There is no evidence that infection with these seasonal coronaviruses results in infertility.

As is standard practice, we asked all women of childbearing age to use effective contraception during our trials, and then tested them before each vaccination to make sure that they were not pregnant. We therefore did not expect any pregnancies in our trial participants, and did not expect to be able to get any information from our trial about the effects of the vaccine on fertility. However, because real life doesn't always go as planned, there have been quite a few pregnancies during the trials. These have been evenly split between the vaccine group and the placebo group. If the vaccine really did affect fertility, we would have seen fewer pregnancies in the vaccine group. The claim that vaccines would affect fertility worried many people, but it is completely false.

If you are going to do your research about vaccines yourself, be prepared to spend a bit of time on it, and be careful where you get your information. Don't inadvertently become part of the 'infodemic' of false and misleading information about Covid-19.[15] Take a look at the Vaccine Knowledge Project website which is run by academics at the University of Oxford and provides independent advice about vaccines and infectious diseases based on the latest research, with no funding from pharmaceutical companies.[16]

—

Entirely separate from the risks and benefits, some people are opposed to some vaccines because of the way they are manufactured, using a cell line derived from human foetal cells. The Oxford AstraZeneca vaccine, for example, like other replication-deficient adenovirus-vectored vaccines, is grown in HEK293 cells. HEK stands for 'human embryonic kidney'. HEK293 cells all originate from the kidneys of a single foetus aborted in the Netherlands in the 1970s, and have been used to manufacture large quantities of many drugs and vaccines that now save lives. This has been of grave concern for the religious right in the United States, who want to ban all research and all use of anything derived from a foetus. It has also been of concern to devout Catholics. The Vatican's position since 2005 has been that it is wrong to make vaccines using a human cell line derived from a foetus, and there is a 'moral duty to continue to fight' against the use of such vaccines and to campaign for alternatives, but that if a population is exposed to considerable dangers to their health then such vaccines may be used on a temporary basis.[17]

And, although only this one legally aborted foetus was used to generate all the billions of HEK293 cells that will ever be needed,* the impression usually given is that abortions are being carried out to order to generate the materials needed. This understandably makes some people uncomfortable but it is not

* This is because cells to which the adenovirus E1 gene has been added are able to grow exponentially and without limit in the lab.

the case, and after decades of deliberate falsehoods and misinformation a strange sequence of events has helped to make that clear.

In September 2020, Ruth Bader Ginsburg died. She had been a liberal judge on the US Supreme Court, and her death gave President Trump the opportunity to nominate an anti-abortion member of the religious right as her replacement.

I am not going to say that abortion is a good thing, but it is sometimes a necessary thing and in a free society a woman should be able to choose if she is going to have an abortion or not. In an ideal world, it would not be necessary for very many abortions ever to be carried out. So an ideal world is what legislators should be looking to create if they want to reduce the number of abortions. Legislators need to stop rape from happening, including the rape of children, rape within families, and rape within marriage. They need to make effective contraception freely available for anybody who wants it, and they need to provide good access to healthcare. They need to educate and empower women to take control of their lives and not have to be subservient to men, including men's views about what they should do with their bodies. The world I want to live in is one in which very few abortions are ever needed because women are not put in the situation where they would want to have an abortion. However, we will never have an ideal world. There will always be mistakes. People will get things wrong, contraception will fail. There will always be some circumstances in which an abortion is the right thing for a woman, and when that is true, then it should be available.

President Trump chose Amy Coney Barrett as his nomination to the Supreme Court, and within days of the event in the Rose Garden of the White House to confirm that nomination, not only President Trump, but many other people who were present that day had been diagnosed with coronavirus.[18] The details of the subsequent treatment that President Trump received raised some questions. We were told that he received oxygen briefly at home when his blood oxygen levels dropped. We know he was then transferred to the Presidential Suite at the Walter Reed Hospital, where a team of doctors took care of him. He was, we were told, given the antiviral drug Remdesivir at an early stage.[19] At the time this had been licensed for emergency use in the United States but its performance had not been particularly impressive. The RECOVERY trial in the UK found that there was some benefit in speeding recovery, but that this was not clinically significant.[20] He was also given another drug that had not yet been licensed and was still undergoing clinical trials.[21] This was a cocktail of two different monoclonal antibodies that bind to the coronavirus spike protein, produced by a company called Regeneron. Monoclonal antibodies are cloned in the lab from a single antibody and they are potentially useful for people who, because of their age or other deficiencies in their immune system, are not making a strong immune response. We then heard that the president was being treated with dexamethasone.[22] This is a cheap and widely available steroid that had been shown in the UK's RECOVERY trial earlier in the year to dampen down the immune response which can be damaging in the second phase of Covid disease.[23]

In the early part of infection with coronavirus, the body is trying to use the immune system to control the virus. But for some patients, there is a second stage of disease where the immune system starts doing more harm than good. And these patients can go into quite a sudden decline. Dexamethasone had been shown to be effective in reducing deaths in people who required oxygen, and particularly in people who were on mechanical ventilation. It had also been shown that it did more harm than good in people who didn't require oxygen. So the news that President Trump was on dexamethasone was interesting. We were being told that he had only required oxygen briefly, and that he wasn't on it anymore. That would mean that he should not have been receiving dexamethasone, which could actually cause him harm. On the other hand, if he was repeatedly requiring supplemental oxygen to maintain his blood oxygen levels, then it would be appropriate to give dexamethasone, but it would mean the disease then was much more severe than anybody was admitting. Dexamethasone also has side effects. It affects mood and it can give a false sense of energy.

When President Trump returned to work he announced that all Americans should have access to the same drugs that he had received, free of charge, including the monoclonal antibodies from Regeneron, which had not yet given any indication of efficacy in clinical trials. (The Food and Drug Administration (FDA) did grant emergency-use authorisation on 21 November, for mild to moderate cases but not for patients who had been hospitalised or required oxygen.[24])

It also became clear that HEK293 cells had been used to test the Regeneron monoclonal antibodies that President Trump had been treated with. Despite this, the religious right explained that this drug was acceptable because President Trump had not required the abortion in question to take place. But the only person that ever required the abortion that led to the generation of the HEK293 cell line was the pregnant woman who made the choice to have an abortion in the 1970s. It was never required by anybody else.*

* HEK293 cells have also been used to assess the effectiveness of Remdesivir, another drug that was given to President Trump.

It is ironic that an event at which President Trump nominated a judge who many hoped would ban abortions in the United States may well have led to him being treated by a monoclonal-antibody cocktail that requires the use of a cell line derived from human foetal cells, which may in turn lead to greater public understanding of how vaccines are made.

It is logical, nonetheless, to ask why we need to carry on using this particular cell line. After all, there are some vaccines that are made using different cell lines. For example, some inactivated viral vaccines are made using Vero cells, which originate from the kidney of an African Green Monkey. The short answer is that switching to a different, non-human cell line is on our wish list, partly because if we did, we could potentially make other improvements to the production process. But it would involve a huge amount of expensive, time-consuming work – including reassessing and recalibrating almost every part of Cath's five steps – that no one, to date, has been prepared to pay for. In an ideal world, we would have started this work ten years ago. In 2020, there was no alternative to using HEK293 cells if we wanted to make a vaccine. On 21 December 2020, the Vatican told Roman Catholics that it was morally acceptable for them to receive Covid-19 vaccines, even if their production employed cell lines drawn from tissues of an aborted foetus.[25]

To return to the subject of vaccine hesitancy: it is natural to be concerned about the unknown or the new, and with drugs and vaccines there is a history that can feed into this concern. However, scientists have learnt from their historic mistakes and put in place measures to prevent them from happening again. Nonetheless, no one can ever assure you that taking a vaccine – or doing anything at all – is completely risk-free. Nothing is completely risk-free. We therefore need to weigh up the risks and benefits when deciding what to do, which is not easy, because we tend not to be very good at understanding risk. Many of us are happy to trust our doctors to guide us. Others prefer to do our own research. Both are reasonable approaches, so long as your information is coming from a reliable source.

Covid-19 is a devastating disease that is killing millions of people and leaving millions of others with long-term health problems. The vaccines being offered have been produced with enormous care and attention to safety; tested to establish the risk–benefit ratio in tens of thousands of people; and will continue to be monitored for rare side effects. Doing nothing is a decision too, and one that can have consequences, just as much as a positive action. Getting yourself vaccinated will protect you and those around you from falling ill with Covid-19, and is the best means we have to bring an end to lurching from lockdown to economic failure via overwhelmed health services. My advice would be to accept a Covid-19 vaccine as soon as you are offered one. That's what I did.

CHAPTER 10

Vogue

Robert Peston, ITV *News at Ten*, 15 July 2020: 'What I've learned is that this very important peer review of the work that's going on at Oxford, backed by the pharmaceutical giant [AstraZeneca], shows response from antibodies and so-called T cells or killer cells that we all have has been as good as the researchers could possibly have hoped. So it is good news, there is evidence that this vaccine may work . . . those close to it think that there could well be [this vaccine] in mass production as soon as the autumn.'[1]

I think I actually spat out my wine. WTF? All of us involved with the project who had had even a whiff of the trial data were under strict instructions not to talk about it. We were not allowed to save copies to our computers, or send it in emails. We were certainly not allowed to tell our friends. There were people who had actually made the vaccine who had not yet been told what the data was looking like. And yet there was Robert Peston on Twitter and on the *national news.*

Articles immediately appeared in *The Times* and the *Telegraph* repeating some of this (other papers were majoring on whether we were going to have to wear masks in Pret or not).[2] The university even tweeted a link to one of them, which seemed like madness: we weren't allowed to talk about our findings but science by leak was OK.[3] I don't think we ever worked out where the leak had come from (speculation was rife: was it someone from the Vaccines Taskforce, or had a government minister wanted a good news story out on a particular day?). But it made us even more careful in the future. And it drove home to us, if we hadn't understood it already, how bright a spotlight was being shone on everything we were doing.

—

When I was I kid I dreamed of being famous: that I would be on the TV and my photo would be in the *Daily Mail*. So I decided to become a scientist. Said nobody, ever.

Throughout 2020, everyone on the team was stepping up to the task of their life. At the same time, because things weren't already hard enough, we were on a steep and sometimes scary learning curve, developing a whole new skill set. Working things out as we went along, we were learning how to navigate the communications challenges of being part of 'the only story in the world'.

We were trying to work out how to communicate complexity when people wanted simple answers. How to maintain patient confidentiality (and avoid insider trading) but also be open and transparent. How to be truthful about our work but not undersell it. How to balance the desire to discuss the caveats and unknowns and uncertainties around our data with the need to reassure and persuade. And we were trying to promote understanding of the science while maintaining some boundaries around our personal lives.

Sometimes the media – traditional and social – helped us to manage these tensions. Like us, most journalists were doing their best in difficult circumstances, and many did an excellent job at communicating emerging and complex science. At other times, it seemed like the media was actually creating the tensions. Sometimes we were frustrated at the way information got distorted or sensationalised in the reporting: some journalists had an agenda, and others did not take care to check the truth and simply presented as fact what another journalist had said. At the same time, this was our opportunity to reach audiences that normally would take no interest in our work, and to get verified and accurate information out into the world. We were always conscious that our founding father Edward Jenner's big achievement was not vaccinating against smallpox. That was not his idea and like most scientists he was building on the work of others. What he did, that others didn't, was publicise his findings and push for the widespread uptake of vaccination.

The media as double-edged sword is not a new idea. But since we were not influencers or members of the royal family, I suppose it was not something to which most of us – working anonymously in our labs and happy if our latest findings got published in a highly specialised scientific journal – had previously given much thought. Anyway, we had no choice but to pick up the sword and wield it as best we could. What we were doing was important and people wanted to know about it. We couldn't choose not to engage even if we wanted to.

Throughout the early months of 2020 there was quite a gradual build-up of attention, pressure and clangers. By the end of January the world was starting to get worried and we were getting a lot of requests for statements. In February, we did our first of several photo shoots in the research labs (not the CBF – I couldn't risk contamination or disruption of the vaccine production process while we were still making it). Having photographers and film crews in the lab was always disruptive. Three or four people would tramp through, creating trip hazards with their trailing cables, dumping their kit on lab benches, and asking stressed people to explain what they were doing, one more time, and then one last time, and then one final time. At least once it ended in tears, such as when a crew, having decided that the office they were supposed to be filming in wasn't suitable because of scaffolding visible through the window, set up to film an interview in the busiest part of the immunology lab and destroyed a whole morning's work.

By March, I and members of my team were getting requests all the time to do press interviews about the vaccine. I remember doing a phone interview with a journalist from the *Express*. I spoke with him for about half an hour and I thought that I had taken him carefully through the story: how we make the vaccine, how we test it for safety, and our plans for the trial. Imagine my surprise when the resulting headline was 'Coronavirus vaccine: Oxford University scientist fears UK "putting eggs in one basket"'.[4] The stupid thing was I had said those words. Right at the end, he had asked me 'what are you most worried about?' This is, now I look back at it, an obvious journalist technique. We'd been chatting, I felt relaxed, so I answered off the cuff. What I meant was (and it is buried in the article, which is actually pretty good: clear and accurate) that because we had to work so fast, Sarah hadn't spent years testing different versions of the vaccine before committing to the final design – we just went for the simple option of repeating the strategy that had been effective for MERS. But that wasn't really what the headline implied. A lesson learnt.

By April there had been reports of paparazzi on campus (I think someone was spotted carrying a camera with telephoto lenses in the hospital car park). We weren't used to this kind of attention. The events around the fake story of the supposed death of Elisa Granato made it clear that we were going to need to be more strategic and controlled about our interactions with the media. The media had been useful to us in various ways. The coverage had probably been a big part of why we were able to recruit to the trial so quickly, and it had also raised our profile with people who ended up generously donating to the project. But now we needed to protect the privacy and security of the volunteers and the trial process, and give ourselves the space to get on with the work.

We put a note on the website we had set up for the vaccine trial, saying 'We are aware there have been and will be rumours and false reports about the progress of the trial. We urge people not to give these any credibility and not to circulate them. We will not be offering a running commentary about the trial but all official updates will appear on this site.' We also started sending all requests from journalists to the university press office, who would decline most of them on our behalf. The press office organised for us to do a few background briefings with selected journalists. These discussed general issues such as the difficulties of comparing the results from two different clinical trials, and how we might deal with that; or how we measured antibodies as part of a clinical trial, rather than giving any kind of running commentary. Other than that, we would communicate when we had data to publish and something to say, and not before.

Of course that did not stop the requests from coming. I was still relatively anonymous, despite my adventures on Twitter, but Sarah had to ditch her phone. Her number was on the university website and she was constantly being called by journalists who wanted a quote, members of the public who wanted reassurance, investment advisers who wanted inside knowledge, and conference organisers who wanted a speaker. She couldn't abandon her email though and so her inbox filled up with sophisticated and not-so-sophisticated phishing attempts trying to get her to click on rogue attachments, and messages from people urging her to drop her research and instead follow up their theory about curing Covid with bananas/garlic/daffodils/Japanese plum extract/plants/all natural ingredients/a compound formula/an antidote to spider poison/water and salt/animals ('I am not specifying the animal'). The university added extra layers of security – there were rumours that Russian agents had been trying to steal data about our vaccine, and we had experienced some 'unusual' problems with our IT. But whatever they did, it certainly did not filter out all the perhaps well-meaning, but ultimately distracting approaches.

Our new strategy – essentially, keeping our heads down until we published our first peer-reviewed paper – was going pretty well until Robert Peston's big reveal. It was a shock, but unlike in April when we were dealing with fake news, at least his summary was correct. The vaccine was doing exactly what we had hoped for. It was fooling the immune system into mounting a defence to the SARS-CoV-2 spike protein and it was not causing any problematic side effects. Occasionally a sore arm, and sometimes an elevated temperature, all easily treated with paracetamol. Just like in April though, the demands for comments went through the roof. This would be the first of many times we were put in an impossible position. Any small piece of information, even if leaked by someone elsewhere, led to us being questioned about it. If we didn't say anything, we were criticised for lack of transparency and failing in our duty to inform the public, and if we did, we were criticised for conducting 'science by press release'.

After that there seemed to be something every month. There were reports in August that President Trump was going to 'fast-track' our vaccine.[5] That created brief mayhem. In September, the trial was stopped because someone became ill and we had to look into whether it was because of the vaccine. Mayhem again. In October, the Russians launched a disinformation campaign, complete with images, memes and videos, aimed at convincing people that our vaccine would turn them into monkeys.[6]

The Science Media Centre – a charity whose philosophy is, brilliantly, 'the media will do science better when scientists do the media better' – became a crucial part of our lives. In normal times they organise events at which scientists with newsworthy findings can speak to journalists and answer their questions. They also advise scientists on how to get their science across clearly and accurately. After lockdown, they started organising online events for us, often at incredibly short notice, and they have been tireless in collecting quotes from experts whenever there is a new development. That has been really important for helping journalists to make sense of it all and to sift out some of the nonsense.

The reporting of the efficacy results from our phase III trial in November illustrated almost every one of the tensions we were managing when trying to communicate about our work. Our results were complex. Unlike when Pfizer and Moderna had announced their results, we had not one simple efficacy number but three: 90% (the group who had had a half-dose then a standard dose), 62% (the group who had had two standard doses), and 70% (both groups together); and the results came from a combination of trials in different countries. Our results, like the results of all the trials, were also market-sensitive. This meant they had to be announced by press release immediately they were known, with the full peer-reviewed academic paper containing all our findings following as quickly as possible afterwards.* Some of

* It is not a coincidence that phase III trial results have tended to be reported on a Monday. Getting a trial data set ready for 'unblinding' takes some time, but once it is ready, the analysis is fairly quick. If the analysis can be done on a Friday evening

those commenting on the situation and accusing us of doing 'science by press release' as a way to get data out without proper scrutiny did not understand this. And until the paper was written up, peer-reviewed and published, we could not defend ourselves beyond saying: just wait, it's coming.

Even once the paper was published, there remained some unknowns, uncertainties and caveats around our data – as there are around all data. But while setting all of that out made everything more transparent – so that's good – focusing on the unknowns and uncertainties also risked undermining confidence in the most important finding: we had a very safe and highly effective vaccine. A particular bugbear of mine – because it involved my work – was the way the issue of the supposed 'dosing error' got written up. We were being bombarded with questions from journalists: what happened? Did you make a mistake? How can we trust your vaccine if you are always making mistakes? Headlines said things like 'clarity needed', 'problems with dosage levels revealed', 'scrutiny grows', 'disquiet mounts', 'questions grow', and 'Oxford vaccine error'.[7] The reality of how some volunteers came to be given a half-dose was complicated, as I described in Chapter 6. The impact of that half-dose on our results was unclear at that point. We had to report the numbers that way because that was what we had pre-agreed we would report, but we needed time, and possibly more data, to understand exactly what they meant. It was your basic scientists' nightmare in a world that wanted simple, clear answers and a good headline. We also maybe did not always explain it well ourselves. But describing it as an error was

after the markets are closed, that leaves two whole days to work out what the analysis means, inform all the people who need to be informed, in the right order, write the press release with quotes from relevant people, and be ready to make the announcement before the markets open on Monday morning.

inaccurate, and implying that we were gung-ho or accident-prone was unfair. It was annoying because it impugned our work and our concern for the safety of our trial participants, but also because it risked undermining public confidence in our vaccine.

Another example of the important and complicated relationship between science and the media was the claim made in a German newspaper in late January 2021, and repeated on Reuters and elsewhere, that our vaccine was only 8% effective in older people.[8] This claim was completely untrue, to the extent that we were all completely baffled about where on earth that number had come from. But the truth took a little time to explain. At that point we did not know what the true figure was. We did not have the data to say: efficacy in over-65s is not 8%, it is x%. This was because we had not brought people over the age of 65 into our trial until we had gathered a lot of safety data on younger people. So by the time our trial reported its results in November 2020, not many of our older volunteers had tested positive for Covid. But while we did not have this clear, simple efficacy data, what we did have was extremely strong evidence, from the immune responses we were seeing in our trials. Since immune responses were similar in younger and older adults, it was to be expected that efficacy would also be similar. Just a few weeks later, we did have evidence showing that, in real life, our vaccine was highly effective not just in the over-65s but in the over-80s.[9] However, by that time Germany's vaccine policymaking body had announced that it would not authorise the use of the vaccine in over-65s.[10] Several months later, the untrue '8% claim' was still being repeated in articles discussing why the vaccine roll-out was going so slowly in Europe.[11] That one bad article, and then the frequent subsequent citing of it, even when accompanied by the caveat that the claim wasn't true, probably did contribute to a general reluctance to take the AstraZeneca vaccine in Germany and elsewhere. Like the kinds of myths we talked about in Chapter 9, it had been thoroughly debunked, but it lingered in people's minds and refused to die. And of course this isn't just about my hurt feelings that my work was misrepresented. It's not even just about the importance of truth in journalism. This kind of misinformation – and this was just one example, although it was a particularly bad one – will have cost lives. People who could have been vaccinated were not vaccinated, and some of those people will have died.

—

We had agreed to make a BBC *Panorama* programme about our year, though we did slightly curse that decision a few times afterwards. However, by December we were pleased that this at least would give us a chance to put our story across accurately. I was also happy to have the opportunity to showcase my part of the story. Manufacturing is not particularly sexy, and with my business head on I have to fund my very expensive facility from projects, so it's important that academic scientists and new biotech companies know we exist and what we can do for them. A mention on *Panorama*, hopefully in a positive light, would be great marketing.

I was very nervous though. I felt the same stresses as when I had done James O'Brien's radio show, and every public-facing interview since. Will I say something daft or wrong? Will he ask me slippery questions, trying to catch me out? Only now there would be cameras and lights as well.

My interview with the BBC's medical editor Fergus Walsh was scheduled for 3 December, at St Cross College in the centre of Oxford. The day dawned cold and wet. I had a lot of Zoom meetings slotted into my calendar for the morning, and I needed to go into the CBF too. I was also in the process of selling my house (a consequence of getting divorced) and needed to drop off some documents with my solicitor. I usually cycle everywhere but so as not to arrive at the filming all sweaty I had booked a car-share for the afternoon. I grabbed my make-up bag and hairbrush, put on a clean jumper – I'm as vain as the next person – and set off.

After a temperature check at the porters' lodge and lots of hand sanitising I was allowed into the college. The film crew were running a bit late themselves so I installed myself in a side room. The room was full of kit: big carrying cases for equipment, odd lighting rigs, the kind of things that you imagine exist on a film set. Not my comfort zone. I had my iPad and plenty of emails to be getting on with though, so I just sat down and tried not to stare at Fergus Walsh who was quietly working in the corner.

The interview seemed to go OK. The last question Fergus asked me was: is your vaccine safe, and would you take it? I said that I didn't expect to be at the front of the queue but I would be encouraging my parents to accept a vaccine – from any manufacturer – as soon as it was offered. And as soon as I was offered one, I would do the same. The point of vaccination was not just to protect me, but to protect the vulnerable in the community, and I would always want to assist with that. Afterwards the team came to do a morning of filming at the CBF. During the actual manufacturing of the vaccine earlier in the year I hadn't allowed any camera crew in, as we have to work in a very clean and regulated way and it isn't easy to sterilise a camera. But now we were between manufacturing campaigns and slightly less busy, so although it was a disruption I gave it the OK.

My daughter Ellie is normally very keen to stay up past nine o'clock, especially if TV is involved. But a couple of weeks later, when I said she was allowed to stay up because Mum was going to be on TV talking about making vaccines, she seemed less than impressed. However, it is testament to the producers that right from the first she was hooked. Not because I was on (*whatever!*). But this was the first time she had seen the numbers of cases and deaths around the world so vividly and clearly portrayed. I think it made her realise how serious what we were doing was, and maybe she felt a little proud. She also loved the footage of the machines filling the doses on the production line and asked whether I had a cool machine like that. As you know, having read Chapter 5, I sadly do not: we do it by hand. I got equally positive responses from family, friends and mums on the school run. Everyone said that they had learned something, and felt reassured about the vaccine.

There was a lot of press coverage in response to the programme too and it all felt positive. The *Daily Mail* described me, to my mum's delight, as having 'a ribald sense of humour and an earthy Kentish accent'.*[12]

—

If we were unprepared for the level of interest the media would show in our work this year, we were even more unprepared – and surprised – by the interest shown in us personally.

At the beginning, there was a lot of interest in us being 'women scientists'. I know that Sarah found this particularly annoying: she is on record as saying something like 'This is

* I know the journalist was entitled to name my accent having seen the programme. But claiming to know my likely choice of jokes seemed an assumption too far (although, I will concede, it might be true).

2020. Why are we even discussing women scientists? I'm not a woman scientist, I'm a scientist and more than half my colleagues are women and we do the job.'[13]

However, I think that with our increased public profile our attitude has become more nuanced. We realise that we are being seen as role models, and that is I think important. Women are still under-represented in so-called STEM (science, technology, engineering and mathematics) subjects at school, at university, at work, and in senior positions at work.[14] There is still a pay gap, and the reasons for that are complex, but it is partly because women are less likely to reach senior positions. There are some things about being a woman in science that are still particularly challenging. Attempts at quick fixes can be counterproductive. For example, in the past deciding to have a senior woman on every committee imposed huge burdens on the few senior women around, preventing them from getting on with their research. But mentorship, allyship and structural changes (particularly around parental leave and fixed-term contracts – which would also benefit our male colleagues) would help. And clearly there is still some work to do on the public perception of women in public roles. It is striking that the male scientists who are equal parts of this project tend not to be introduced in terms like 'Irish brunette mother of two' (Tess), 'serious redhead mother to triplets' (Sarah) or 'not your stereotypical Oxford boffin' (me). I am pretty sure Andy has never been described as 'male scientist Andy Pollard'. For the record, Andy's hair is grey, Adrian's is strawberry blond and Sandy is a brunette.

So while Sarah's initial response – 'why are we still talking about this?' – is valid, the reality is that we should still be talking about this – and not just talking about it, but actually doing something about it – until we don't have to anymore. It is an honour to be held up as a role model, and if our position in the public gaze means more young people consider options that otherwise they would not have done, that can only be a good thing.

There were also lots of documentary makers who all wanted to film us riding our bikes. Because it's the cliché of the Oxford academic isn't it? (I admit, in this case it's a cliché that is true: both of us do cycle to work.) Sarah told me that the trick, which she had learned the hard way, was not to let them persuade you to cycle down the big hill next to the campus, because then you had to get back up again. Also, they liked filming us walking through doors, going downstairs, and opening freezers. We became quite practised at it.

One of the more surreal moments of this year was Sarah's high-fashion photoshoot in the basement of the Jenner Institute. Sarah had already been to a London studio for a shoot for *Vogue*'s '25 Women Shaping 2020' earlier in the year.[15] She said she thought it would be fun, and not something she was likely to be asked to do ever again. But not long after *Vogue*, she was approached by *Harper's Bazaar*.[16] She describes their shoot in the basement, which was part of their Women of the Year series, as 'all slightly ridiculous – the sleeves of the jacket were too long which is why my arms are in such strange positions'. Although it *was* slightly ridiculous, the magazine was making an important point in highlighting Sarah's achievements, in a forum where scientists would not normally get exposure or recognition. (And I suspect she slightly enjoyed it.)

There were other 'pinch yourself' moments throughout the year. The line between reality and fiction almost disappeared for me one evening in May when I sat down to watch TV. I had always loved Charlie Brooker's brand of darkly satirical comedy and thought his *Antiviral Wipe* would be a welcome distraction. Then up popped Andy Pollard, being interviewed by Philomena Cunk. He hadn't told me he was doing it, and it was such a shock – he has a very serious and professional air at work, but this gave away his secret sense of humour. My WhatsApp messages to Tess allude to laughing so much we wet ourselves.

We have sometimes felt that the media attention was too much, and that we would give anything just to return to our previous, unobserved lives. But stepping back, I think that even if us putting ourselves out there for examination has increased the understanding of just a small number of people, it will have been worth it. The hassle and the intrusion and the lost mornings will have been worth it if there is one more person who can see through conspiracy theories, and who might then be confident to persuade a frail elderly person that it's OK when their vaccination date comes round; one kid inspired to do triple science GCSEs meaning one more potential future scientist; one girl brave enough to put her hand up in class to ask a question; one more volunteer for a trial; one more donor to the cause. Maybe what I have really taken on board is that science itself needs to be seen. Then it will be understood, then it will be trusted. Not blindly – some scientists make mistakes, and some cheat – but with robust challenge. And this engagement and understanding will make for better policy, better decision-making and hopefully better futures.

CHAPTER 11

Waiting

1 August–8 November 2020
Confirmed cases: 17.85 million–50.56 million
Confirmed deaths: 680,961–1.26 million

In the autumn we had an infestation of wasps. The nest had been there in a side wall of our house all summer, but it was out of the way, so we had left the wasps to themselves. One Saturday in early October it rained heavily all day. The wasps decided not to venture out into the rain and instead found a route through the wall into our kitchen, and the bedroom and shower room above it. We couldn't see where they were getting through but they kept appearing.

I remember the first time I was stung by a wasp. My father always insisted that wasps and bees were not interested in stinging us and that, provided we did not panic and flap about, they would ignore us. Unfortunately, not all of the wasps seemed to play by the rules. Aged around 7, I was lying on a camp bed in a tent in our back garden, quietly reading, when a wasp announced its presence by stinging me on the leg. I ran into the house, crying not just from the pain but also from the injustice. I most certainly had not been doing anything to attract its attention, but had been stung anyway. I was given a cold wet flannel to put over the sting and sent back out into the garden. (My mother's style of parenting required that we should entertain ourselves when at home, preferably outside, not in our bedrooms, and most definitely not 'under my feet'.)

Despite having been stung I was still not particularly afraid of wasps. We all got ant stings repeatedly during the summer while playing in the garden and the wasp sting wasn't so much worse. Later that year though, or possibly the next, my mother suffered an allergic reaction to a wasp sting. The wasp had landed on the back of the kitchen tap and she didn't see it as she grasped hold of the tap firmly to give it the wrench that was necessary to turn it on. The wasp, trapped, gave her a sting that went deep into her finger. Shortly afterwards she began to feel unwell. My brother, my sister and I were all in the garden (as per her request) but fortunately my brother was within earshot, and was dispatched to fetch our strong and capable neighbour Mrs Mansfield. My father arrived home for his lunch (though we called it dinner) to see Mrs Mansfield carrying my mother out to the waiting ambulance.

There were no lasting physical effects of this encounter, but whenever my mother saw a wasp after that she would engage in the panicking and flapping about that was forbidden to the rest of the family, repeating in a high-pitched voice 'Oh, wasp! I'm allergic!' Unfortunately, she passed on the panic response to my children when they were at an impressionable age, along with 'Oh, wasp! Grandma's allergic!' So when the wasps began invading our house, the children were far from impressed.

On that wet Saturday we had about thirty wasps in the kitchen. Over the next week we counted about ten new wasps a day. Some turned up dead on the windowsill, some waited quietly to be scooped up and ejected, others buzzed about inside the lampshades. I was reluctant to call pest control because any poison sprayed into the nest would have made its way into the house. Instead we kept doors shut, and removed the intruders with as little fuss as possible. Occasionally, their incessant buzzing did get to me though. A week after they first broke through into the house, Rob had gone out to fetch fish and chips for supper. I was setting the table in the kitchen when I saw a dead wasp on the windowsill. As I removed it, I found two more, either dead or comatose, trapped inside the blind, and fetched the hand-held vacuum cleaner to remove them. Then the buzzing inside the lampshade started up. As Rob pulled up in the driveway I was dispatching the last of three more wasps with a vacuum cleaner in one hand and a fly swatter in the other, using a badminton-style forehand to swat and the vacuum cleaner to collect.

Over the summer and through the autumn, as the UK gradually opened up and closed back down again, and the wasps went from being something outside that didn't bother us to something inside that did, we were waiting. We all knew that every day might be the day we could unblind our trial and find out whether and how well our vaccine worked. In the meantime, the US trial got caught up in politics. We mostly managed to look on all the buzzing as harmless irritation that we should just ignore. But occasionally it got to us. And sometimes, even when you are not flapping about, you get stung anyway. In the end, all we could do, with the wasps and the trial, was wait, with as much patience as possible, for this to pass.

—

All year, I was constantly being asked, by journalists, by friends, by colleagues: when will we know? When will we know whether the vaccine is safe? When will we know whether it works? I was always confident, because of the work we had done on similar vaccines before, that it would be very safe, and in July we had published the data that demonstrated that. I was also pretty confident from the outset that it would work, but until enough people in our trials had tested positive for Covid, and the trial was unblinded, we couldn't know for sure.

Unfortunately for the many people, including me, who were desperate for information, it wasn't possible to keep a running score. The problem with doing so is that it is then tempting to choose to end the trial at a point where the numbers favour the outcome you are hoping for. Imagine that the first positive case was in someone who had received the ChAdOx1 vaccine (the vaccine group). At that point, the vaccine looks ineffective. Then imagine the next three cases were in people who had received the placebo injection (the placebo group). Now the vaccine looks much better. If the next six cases were all in the placebo group, the vaccine would look really good. It would be tempting to stop the trial there. But what if the next ten cases were all in the vaccine group? We would have stopped the trial on a vaccine that was now looking no better than a placebo. So, it is important for the scientific integrity of the trial that there is no opportunity for those conducting it to know how things are going.

Added to that uncertainty, over the summer and the early autumn, we and AstraZeneca were having lots of discussions with multiple regulatory authorities – the MHRA in the UK, the FDA in the US, and several others – about exactly how many positive Covid cases there needed to be in our trial before we could unblind it. The more cases in the trial before we unblind it, the more confidence we can have in its results. On the other hand, the more cases we needed, the longer it was going to take and in the meantime people were dying.

Another discussion we were all having with the regulators was whether it was going to be acceptable to pool results from trials being carried out in different countries. It's a good thing to have data from different countries for a vaccine that we hope will be used worldwide, but the different patterns of the outbreaks in different countries meant that we might see different levels of efficacy reported. This was because the vaccine was likely to be – and did turn out to be – very effective against hospitalisation and severe disease but less effective against mild disease. In a country with high levels of severe disease, the vaccine would therefore look more effective. This was illustrated very clearly later on when another vaccine manufacturer announced efficacy of 78%, then modified that to say it was 50% when taking all mild disease into account.[1]

It was also unclear, at this point, what might happen when the first vaccine – ours or someone else's – reported its trial results. Almost every vaccine under development was based on the same spike protein. Many had already published the data from their phase I safety trials, showing that the vaccines were well tolerated, only resulting in the kind of things we would expect to see after receiving a vaccine (short-lived pain at the injection site, fatigue and so on). Some were also reporting the data from their phase II immune response trials, showing that they were inducing neutralising antibodies against SARS-CoV-2. But what we didn't know was, what *level* of antibody response we needed to protect against disease. What the phase III trials were trying to establish was, primarily, whether the immune response induced by a given vaccine was protective: did the antibodies stop people exposed to Covid-19 from getting ill? We were also very interested in other questions, like, did the vaccine prevent asymptomatic infection (because if it did, that would be really important in reducing virus transmission).

If we were able to compare the immune responses of different vaccines that were in phase III trials, could we assume once one vaccine had been shown to be protective that this meant any vaccine that induced a comparable immune response would be protective as well? Might it make sense to unblind a trial early in those circumstances, rather than waiting another month or two months for it to gather enough cases?*

So, as the summer turned to autumn and the days grew shorter, only a small group of statisticians working on the project actually knew how many positive cases we had accumulated in our trial. And even they didn't know how many of those had received our vaccine and how many the placebo. For everyone else it was a macabre waiting game, played in the dark. We had to agree how many cases in which trials would be needed before we unblinded, and under what circumstances to combine data from trials in different countries. Then we just had to wait for that number of people to fall ill and test positive for Covid.

In August, with 15,000 people enrolled and vaccinated into phase III trials in the UK and Brazil and a smaller phase I/II study in South Africa (plus another separate trial underway in India) we knew that it could in theory be any day, but thought it would probably not be before October. Transmission in the UK had dropped to a very low level since the March lockdown, which was great for the population as a whole but a problem for our trial.

Right at the end of August, the AstraZeneca-led US trial finally got underway. It was scheduled to recruit 30,000 people very fast, in a country where case numbers were high and rising.

---

* In the event, this last uncertainty became a moot point. Assay (i.e. test) development had not kept up with vaccine development and there were no validated and standard assays that would allow us to reliably 'score' and compare the complicated immune responses induced by different vaccines.

The data from this trial would not be pooled with the data from the UK/Brazil/South Africa trials, because it was being run with a slightly different protocol. However, for licensure in the US, it was going to be important to get efficacy data from the US trial. And then, almost as soon as it had started, it came to a halt again.

The US government under President Trump had from the outset stood apart from the rest of the world on its approach to tackling Covid-19. Despite repeated warnings from scientists and international organisations that 'no one is safe until we are all safe', and despite the obvious fact that viruses do not understand or respect international borders, the US had adopted an America First stance. They withdrew from the WHO and refused to join Covax, the global collaborative effort spearheaded by WHO, CEPI and Gavi to speed up vaccine development and ensure fair access to supplies, especially for low- and middle-income countries.

By May, while elsewhere infection rates were falling and lockdowns cautiously being lifted, the situation in the US was bleak. Case numbers had risen precipitously in late March, making the US officially the worst-hit country in the world. It had suffered over 2,000 deaths a day throughout most of April and by the end of May the official death toll had passed 100,000. In one week in May the case numbers had quadrupled. To accelerate its efforts in the fight against Covid, which were being heavily criticised as too little too late, the Trump administration had created a new public-private partnership called Operation Warp Speed (OWS).[2] Clinical trial oversight and emergency-use authorisations would still come from the FDA and funding would still come from the Biomedical Advanced Research and Development Authority (BARDA). OWS was supposed to make everything happen fast: one of its main goals was to provide

300 million doses of Covid-19 vaccine for the United States by January 2021.

I had had many dealings with BARDA over the years through my work on influenza and emerging pathogens and I had not gained the impression that it was an organisation well suited to operating at warp speed. When I had visited its headquarters in Washington DC to give a seminar on influenza vaccine development, I had spent most of the day traipsing in and out of buildings, each time going through airport-style security measures. Coffee had to be bought from a local Starbucks outside the building because BARDA staff could not accept anything that might be considered a bribe, nor give away anything for free. As I was not a US citizen, I was not allowed to give a talk in the BARDA building so out through security we went again, as did most of the people who had turned up to listen to me.

Around the time that OWS was announced, Rick Bright, BARDA's director, was 'reassigned' to another post, and Moncef Slaoui was appointed as head of OWS.[3] Rick Bright had been director of BARDA since 2016 and with years of experience in the field was well qualified to lead its pandemic response. He had been frustrated by the lack of attention being paid to SARS-CoV-2 in the US, and then had refused to back hydroxychloroquine as a treatment for Covid-19 in the absence of trial data and FDA approval.[*4] Moncef Slaoui also had a long history of working in vaccine development, although he had never

* The story of hydroxychloroquine, a cheap and readily available antiviral drug, is complicated. President Trump repeatedly championed it, directly contradicting guidance from the country's top public health bodies. It was not necessarily the case that hydroxychloroquine provided no benefits in treating Covid-19. Rick Bright's position, which was correct, was that before rolling it out, clinical trial data was needed to establish whether it was effective, in what circumstances, and with what possible harmful side effects.

worked on vaccines for emerging pathogens. At the time of his appointment he resigned from his position on the board of US company Moderna, who were developing an mRNA vaccine against SARS-CoV-2, but retained stock options worth $10 million and a seat on a board of Lonza, who had partnered with Moderna to produce the vaccine. Following criticism about his conflict of interest, he later gave up the stock options and resigned from Lonza and other positions.[5]

OWS awarded $1.2 billion of funding to our vaccine almost immediately, but from there it took some months to get the US trial started, with the FDA taking things rather more slowly than the MHRA had done. In the meantime, President Trump, hoping for re-election that November, was taking the line that everything was going to be fine because a vaccine would be here to save everyone very soon. Scientists, health professionals and regulators became concerned about the political pressure being brought to bear on the FDA to improperly rush a vaccine through. FDA directors made it clear that they would resign if pressured to approve a vaccine before it had been properly tested.

Some vaccines whose trials had started later than ours had not yet amassed very much safety data. No regulator was going to license any vaccine until the safety data had accumulated. We *did* have a great deal of safety data accumulated for our vaccine: we had already vaccinated thousands of people and most vaccine-related adverse events occur very soon after vaccination. But we still needed to continue with our phase III trials in order to establish how effective the vaccine was going to be. In particular, we needed to establish how effective it was going to be in older people – who as we know are more at risk from Covid-19 and who often produce weaker immune responses to vaccines. There was also a concern that rushing the vaccine approval, even for a vaccine like ours where the trial had already amassed enormous amounts of safety data, could put people off being vaccinated, either with this vaccine, or with other important vaccines like MMR. It was, almost everyone working in the field felt, undesirable to have a situation in which a vaccine had been rushed to licensure because a politician wanted to get re-elected.

On 25 August the *Financial Times* reported that President Trump was considering 'bypassing normal standards' and fast-tracking approval of the Oxford AstraZeneca vaccine so that it could be rolled out before the country went to the polls in early November.[6] Our phase III trials started in the United States less than a week later. It wasn't great timing because it came just before the Labor Day holiday so lots of trial centres decided not to get going until after the long weekend. By the time America came back to work after the long weekend, however, our trials had been paused.

There had been what we call a 'SUSAR'. A SUSAR is a *suspected unexpected serious adverse reaction* in a clinical trial. An *adverse* event is something that happens after taking a drug or receiving a vaccine, which may or may not have been caused by the drug or vaccine. Everything has to be recorded, even if it is pretty obvious that the vaccine was not the cause. A *serious* adverse event is one that results in death, is life-threatening, results in long-term or significant disability, or is bad enough to require a person to be hospitalised.[7] For every serious adverse event, it is necessary to assess whether it was caused by the vaccine or not, and the outcome of that assessment may be that it was not associated, possibly, probably or definitely associated, depending on the level of certainty of the clinicians who make the assessment. An adverse event that is caused by the vaccine is known as an *adverse reaction*. In the clinical trial protocol for a particular vaccine, there will be a list of expected adverse reactions. For example, having a sore arm is a known adverse reaction to getting a vaccination. An *unexpected* serious adverse reaction is one that we didn't previously know was associated with the drug or vaccine in question. And a *suspected* unexpected serious adverse reaction, which is the SUSAR, is something that might be an unexpected serious adverse reaction, or it might not. It might just be something bad that would have happened anyway (an adverse event unrelated to vaccination) and just happened to occur shortly after taking the drug or vaccine.

Sometimes it is immediately very clear that a serious adverse event had nothing to do with the vaccine and so is not a SUSAR. One volunteer on Moderna's Covid vaccine trial got struck by lightning.[8] This had to be recorded as a serious adverse event, but no one thought it had anything to do with the vaccination.

In other cases things are less clear and have to be investigated. There was a case early on in its roll-out in which an apparently healthy teenage girl very sadly died hours after receiving an HPV vaccine, which protects against cervical cancer. After investigation it was found that she had had a very large, very rare tumour, which no one had known about, and it was this that caused her death, not the vaccination, which she just happened to have on the day she died.[9] Or, imagine 80-year-old Mr Smith going to his local pharmacy for his routine flu vaccination. When he gets there, though, and before he has a chance to have the vaccine he starts to feel unwell. The pharmacist recognises signs of a stroke and calls an ambulance. While he is being treated, Mr Smith's wife calls several friends, very upset, and warns them not to have the flu vaccine because it has just caused her husband to have a stroke. The story gets picked up by the local papers and it is only when Mr Smith's condition has stabilised that he is able to tell his wife that the stroke was nothing to do with the vaccine because he hadn't actually received it. But had the vaccination gone ahead, Mr Smith would have had a vaccine and then, immediately afterwards, a stroke.

The point is, sometimes things that happen after a vaccine is given are nothing to do with the vaccine. They happen around the same time, but that is a coincidence. The one thing has not caused the other. So, a SUSAR is a serious event that might or might not be an adverse reaction to a drug or vaccine during a clinical trial, and that needs to be investigated. It could have been caused by the vaccine, or it could just be something that happened around the same time. Any trial in which a SUSAR has taken place may be put on a voluntary pause by the chief investigator (which in this case was Andy), or sometimes a 'clinical hold' will be imposed by the regulator: in both cases, volunteers who have already been vaccinated will continue to be followed up but no further vaccinations will be given until the investigation is finished. We wouldn't want to be vaccinating more people who could then potentially go on to have the same adverse reaction. Sometimes the diagnosis is simple and sometimes it is more complex, as illnesses can be. (Remember it took eight months for me to get a diagnosis of bile acid malabsorption. If I had been part of a trial and this had been a SUSAR I would have got the tests more quickly, but it still would have taken several weeks.)

This was not the trial's first SUSAR. SUSARs are a frequent and normal occurrence in large phase III trials because the more people involved in a trial, the more likely it is that one of those people will fall ill quite by chance. We were also recruiting older people into the trial, and we know that older people have a higher risk of illness than younger people. By this time we had 18,000 people – essentially, a small town's worth of people – in the trial worldwide, and in any small town, over a period of weeks and months, people are going to get ill. Reporting a SUSAR was actually a sign of good safety oversight: it implied the investigators were doing their job properly.

The trial's first SUSAR had been in July, when we had gone through the correct process of pausing the trial, investigating the situation, and presenting the evidence to independent assessors and the regulators, who carried out their own independent reviews and decided that the adverse event had not been caused by vaccination. The trial had then restarted without fanfare or fuss. The expectation was that we would go through a similar process now. The investigation would last for a few days, and then it was likely that the trial would start up again.

But whereas the media hadn't noticed the first time the trial had stopped in July, the second time, they blew it up into a big story. There were journalists hounding every member of the Oxford team. What had happened? Who did it happen to? How are they now? What does this mean for this vaccine and other vaccines? A lot of what they wanted, though, was still being investigated. A trial volunteer had become unwell, and the details were confidential between that individual and the team looking after them, meaning the principal investigator at the trial site and the other clinical professionals brought in to care for them and try to get them a diagnosis and full return to health. The university put out a statement that simply said there was a pause (no more people to be vaccinated), due to a SUSAR that was being investigated – which was the right thing to say – and left it at that. We simply didn't have immediate answers. This was a case where true transparency meant acknowledging uncertainty. In the media, conclusions were jumped to very readily: 'Is this the end for hopes of an early breakthrough?'

said the *Guardian*, while the *Sun* announced, 'Vaccine blow: coronavirus trial put on hold in major setback'. Reuters said that 'suspension of global trials . . . cast a doubt on prospects for an early roll-out of one of the most advanced Covid-19 vaccines in development'.[10] None of this was correct but in some ways the potential implications were more far-reaching than these headlines implied; if this was indeed a genuine adverse reaction to this vaccine, then it might be because of the spike protein that all of the vaccine developers were using to induce an immune response. In which case, all vaccines in development would be affected.

—

2020 was in many ways the year that my whole career had been building towards, but there were days when I woke up and wished it wasn't happening. I was very aware of how fortunate I was to have a secure job (finally) and my children all safely at home with me. Yes they were all grown-ups now and I probably shouldn't worry about them, but it was still comforting to know where they were and that they would be looked after if they became ill. I was pleased not to have to travel for work so much, and Rob and I were going for a long walk with friends every weekend so I still had almost as much social life as before the pandemic. But I missed being able to take a whole day off without checking in to see what new problem had come up. I missed having some control over what I worked on. And I was very tired of talking endlessly to the media and always having to be the 'adult in the room'.

In fact, while the trial was on hold and this latest media storm was raging, my son, who like my daughters had been studying at home since March, did go back to university to start his new term. I was pleased for him but a bit sad for myself – I would miss him. He is very sensible and doesn't need me fretting over him, but, as many parents will know, once you have had to pick up a child from a pool of blood after his ill-judged attempt to rollerblade down a bumpy pavement; or found out several months later about a bus crash in Quito ('You remember I told you about it'), it is hard not to conclude that you will only truly know that that child is safe when he is actually visible to you. It was probably a good thing I had work to distract me when he left.

In the event, the investigation was completed and the UK regulator gave the go-ahead to restart six days after the trial had paused. This unleashed yet more questions from journalists: why has it only partially restarted? Why has it not restarted in the US? The answer was that this was a developing situation and each regulatory authority had to make its own decision. Brazil restarted two days later, and South Africa three days after that. Although someone could always be found to give a quote along the lines of 'it takes five to ten years to make a vaccine, it isn't safe to rush', most of the coverage did become more sensible, explaining that a pause to investigate illness in a volunteer was a perfectly routine part of a clinical trial.

Conveniently, the UK trial resumed a couple of days after a group of nine major pharmaceutical companies, including AstraZeneca, jointly signed a declaration pledging that, regardless of pressure from politicians, none of them would submit a vaccine for approval before it had been fully tested for safety and efficacy, as determined by the relevant regulatory authorities (in the US this was the FDA).[11] This gave commentators an opportunity to point out that our pause had been a demonstration of good clinical trial oversight and careful attention to the safety of the volunteers in the trial. We had always said that we would go quickly but prioritise safety and the fact that we had been prepared to pause the trials across the world to investigate a participant's illness was seen as a good thing. The timing of our SUSAR was chance of course. But the timing of the declaration, aimed at ensuring public confidence in the science and ethics behind Covid-19 vaccines, just a few days after it had been reported that President Trump wanted to fast-track a vaccine to roll it out before election day, was not.

The FDA response was to publish new guidance in early October on its requirements for approving any vaccine, including a requirement for follow-up data for a median of at least two months after final vaccinations in phase III trials.[12] That effectively put an end to any chance of a vaccine in a US trial being licensed before the election in early November. No US trial would have two months of safety data on enough participants by then. However, there was still the possibility that the Oxford AstraZeneca vaccine could be licensed in the UK by the MHRA. And we had started our UK clinical trials in April and continued through the summer so we had more than two months of safety data on large numbers of people. There was a possible scenario in which our UK trial reported its efficacy results, and that trial met the FDA's two months' safety data criterion, and the FDA decided it was happy to review the vaccine on that basis. This seemed to lead to some further concern in the US. Over the weekend, the speaker of the House of Representatives Nancy Pelosi put out a statement saying that she would not trust the UK regulator, the MHRA, to license a vaccine because their procedures were not as stringent as those of the FDA.[13]

This was unhelpful and unfair. From my experience of dealing with them both, I did know that the two regulators worked in different ways. The FDA approach was more process-driven, whereas the MHRA's approach was more interactive, and more focused on gathering the evidence needed to assess the risks and answer the scientific questions. By way of illustration, many years previously we had been asked to collaborate with a US group working on malaria vaccine development. We had already completed a phase I clinical trial on a vaccine. It had been well tolerated, but – as happens a lot in vaccine development – the immune response was not as high as we had hoped, and we were not planning to proceed any further with it. But we did still have some of the batch left and the US group wanted to do a trial using our vaccine in combination with another one to see if that might improve the immune response. The issue we came up against was that although we had completed a clinical trial successfully in the UK, the FDA required toxicology studies to have been completed in two different species whereas in the UK we only have to complete a toxicology study in one species. We had done that and proceeded to human trials, and shown no safety concerns. On a call with the FDA, we explained that we had safety data from mice, and also from humans, which are a species after all, so would that work for them? The answer was no. They needed toxicology studies from another animal species – a rat or a rabbit.

The problem was that if we did a toxicology study in rats or rabbits, it would use up the limited amount of vaccine that was remaining, and we wouldn't then be able to do the clinical trial. We were stuck. If we did the toxicology study the FDA required, we wouldn't be able to do the clinical trial, and if we didn't do the toxicology study they required, we wouldn't be able to do the clinical trial.

My colleague Professor Adrian Hill who was leading the malaria vaccine programme came up with what I thought was a very pragmatic solution. He pointed out that normally a toxicology study would be done in equal numbers of male and female animals to prepare for doing a phase I safety trial in male and female humans. He suggested that if we only did half of the toxicology study, in either males or females, we would only use half of the remaining vaccine. We would then have enough vaccine to do a phase I trial in whatever sex of humans the toxicology study had been done in. It was not something we would normally propose but it would untie the knot.

We were, however, unable to come to an agreement so the clinical trial was never carried out. It was frustrating. But regulatory authorities are complex organisations, whose operations have developed independently out of the needs of the countries in which they operate. They all have very similar aims. They each pursue those aims with integrity, stringency and rigour. But their ways of working and precise requirements can be different. During the pandemic, we wanted the vaccine to be made available in as many countries as possible, which meant engaging with a multitude of regulatory authorities and their different requirements.*

As our trials continued in the UK, Brazil, South Africa, India, and Japan, they remained on hold in the US. Apparently, it is very rare for the FDA to impose a clinical hold on a trial that has already been paused by the trial sponsor – in this case AstraZeneca – and it can be a slow procedure to get it lifted.

October rolled on. Case numbers in the UK were on the rise, especially in university towns and the north, and a new tiering system of escalating restrictions came into force. Despite the high stakes, the mundane day-to-day continued: we still had wasps in the shower room and (in an unrelated incident) our dishwasher broke. Finally, on 23 October, after seven weeks

* The pandemic has reinforced calls for regulatory harmonisation, which would certainly be very welcome from our point of view. But when different organisations have their own ways of doing things it is not simple to get them to agree to settle on one system for all.

of delay, the US trial restarted. When it had been paused, 829 people had been enrolled at nineteen separate sites. Six days after the hold was lifted, that number was already 2,400 and recruitment continued rapidly, reaching a total of 32,000 volunteers, mainly in the United States, but also in Chile and Peru. Trial participants were ethnically diverse; about 20% were older than 65; and 60% had conditions such as diabetes, severe obesity or heart disease that put them at increased risk of severe Covid. Because of differences in trial design (positive Covid cases were defined slightly differently, for example, and the placebo was saline rather than the meningitis vaccine) the US trial data would be analysed separately from the data from the Oxford-led UK, South Africa and Brazil trials. However since it was in itself a large trial, which had started later and would therefore likely reach its conclusion later, we hoped that it would not only allow us to achieve emergency-use approval in the United States, but also provide independent confirmation of the results from the Oxford-led trials.

By the end of October, with case numbers continuing to rise, we felt like we were winning against the wasps. We were down to one or two per day. In quiet moments my mind started to turn to Christmas, and how to safely gather in my son and mother-in-law, who was over 80 and lived on her own over an hour's drive away. However, quiet moments were not easy to come by. Rising case numbers were bad for the country but good for the trial. On Hallowe'en a second month-long lockdown was announced to try to get the case numbers down. With more people across the UK getting infected, and plenty of infections in Brazil, South Africa and the US as well, the day we had been waiting for since the summer was surely getting closer. We were confident we would have results before the end of the year.

Whenever we did get the results, though, there would still be a lot to do before the vaccine could start going into people's arms. We would need to provide all the data to the regulators, who would need to review it and decide whether to license the vaccine for emergency use, and in which parts of the population. Usually a trial would complete all the expected follow-up, perhaps a year for all participants, report its efficacy results, and then the task of preparing the data for the regulators to review would begin. But, as we had throughout the year, we were running tasks in parallel that would usually be done in sequence. We did not know exactly when we would need to have all the analysis ready but, as if preparing to take an exam to be sat on an unknown date, or running a race with an unknown finish line, we were working hard to prepare. There were about half a million pages of data to check. It was not a small task.

Every interaction between a volunteer and a clinic (of which there were tens of thousands) had to be reviewed and the data extracted. Every report of a sore arm or a headache had to be reported in a standardised manner. Every immune response of every volunteer tested following vaccination had to be logged and checked.

There was also a complex process in which an independent adjudication panel had to review every positive Covid case to decide whether it qualified to be included in the primary (the most important) analysis. When someone in the trial developed symptoms that might be Covid, they had to contact the clinic where they had been vaccinated and arrange to go in for a test. At the clinic, a doctor would record their symptoms and do a swab test. The volunteer could then leave, with instructions to follow whatever local guidance was in place at the time for those waiting for a test result. However, this was not enough for someone to be counted as a 'case'. The adjudication panel had to review the record of symptoms to check that they included at least one of cough, shortness of breath, fever above 37.8 degrees, and loss of smell or taste. The Covid test result had to be a clear positive using a recognised PCR test, and it had to have occurred at least fifteen days after the second vaccination.*

Additionally, for the primary analysis, the 'case' only counted if the volunteer had been seronegative (meaning they didn't already have antibodies to SARS-CoV-2) when vaccinated. When our trials had started there had been no way of checking this because there were no antibody tests. This was fine, because we needed to collect some safety data in people who were seropositive, but since they couldn't be included in the efficacy analysis we didn't want too many of them. Later, the staff in our immunology lab developed a test we could use in the lab. We began to test blood samples from the volunteers before we vaccinated them, and only accepted those with a negative antibody result into the trial. As the trials expanded around the country, some sent us blood samples to test for antibodies and others did their own tests. The tests were a bit of a mixture, but they allowed us to get on with the trial using volunteers who were probably seronegative. However, for the regulatory submission, this wasn't good enough. Blood samples that we had stored from the day of vaccination in all the volunteers who had gone on to develop Covid symptoms and have a positive PCR test had to be found from our many freezers and sent

* What made things slightly more complicated was that many of our UK participants had been sending in nasal swabs to be tested every week, so that we could see whether they were developing asymptomatic infection. If we picked up an infection as part of that process, and they then developed symptoms and took another test, the date of the first test 'counted' for the fifteen-days-after-second-vaccination rule. This ruled out some cases that could otherwise have been included in our analysis.

to a lab in the US to be tested again. Shipments of those samples had to include temperature trackers so that we knew they hadn't thawed out in transit. The results were then sent back by secure data transfer to the statisticians, who matched them up with that particular volunteer's other information, for the adjudication panel to review.

Although I knew that all of this was going on, I was not personally involved with it. It was a strange time, waiting for the cases to mount up and be judged so that we could do the analysis and work out how effective our vaccine was. It was in some ways the goal we had been working towards all year. But we also knew it would be the beginning of something much bigger. We didn't know when it would happen, and we didn't know what would happen next. We just had to carry on working and carry on waiting.

—

Sunday 8 November was Remembrance Sunday and the village church was ringing the bells for the service. I had not slept well that week and by the weekend I was very tired and not feeling very productive. Non-work issues were getting to me more than work: the US elections, for one. Like many others I expect, I found myself more than once waking and checking the news in the middle of the night. I don't normally take much interest in the politics of the US any more than I would the politics of other countries that I don't live in. But this year, with the pandemic raging out of control across the US, causing 1,000 deaths every day, the outcome of the election meant a lot more than usual.

Another source of stress for the last two weeks had been my daughters taking their postponed final exams for their degrees.

It was much worse for them than for me of course, but some of the stress rubbed off, and I was concerned that our home Internet might let them down when they needed to download or upload the exam papers. Thoughts about how to get the family together for Christmas were also on my mind.

On Friday I was too tired to work well and I spent Saturday in the office catching up on a lot of things I should have already done. On Sunday I decided to give myself a day off, before returning to the fray on Monday. I did the last day of the thirty-day upper body challenge I had been doing (it had taken me a lot longer than thirty days because I kept adding in unscheduled rest days, but I was determined to finish) and then sat down with a coffee. After that, my mind spun off into a pattern I was familiar with from when my children were babies and I had no free time. It goes – if I had, or now that I have, a few free hours, what would I, what shall I do with them? There was so much I had neglected – what would it be? My garden needed attention, as did my hair and nails. I wanted to be creative, to cook, knit or sew, but all of those required some preparation and I didn't have enough time for that. Maybe I should just try to get some more sleep, or light a fire and find a film to watch with a cup of tea. I might not have another day off for a long time, so how should I best make use of this one? In the end I recognised the pattern for what it was, put on a wash and then forced myself out for a run.

I was still cycling to work but I hadn't been out running since the summer, when I had struggled with my fitness. It wasn't great but it wasn't terrible either and in the last stretch, which is a gentle downhill, I did feel that I was getting into my stride. I reached my front door and was bending down to undo my running shoes, starting to feel very pleased with myself for a good effort that morning, when I heard a buzzing to my right and then felt a sharp jab of pain on my ear. I had not even been allowed five seconds of self-congratulation before the wasp intervened. I felt robbed. But there was nothing to do but head back into the house, ear throbbing.

CHAPTER 12

To Licensure and Beyond

9 November 2020–27 January 2021
Confirmed cases: 51.06 million–100.93 million
Confirmed deaths: 1.27 million–2.18 million

On 8 November 2020, after days of knife-edge uncertainty, it became clear that Joe Biden had won the US presidential election. The following day, more fog lifted as BioNTech and Pfizer announced the efficacy results of their phase III trials. As they had to, they reported their results by press release and there would be a lot more detail to come. Nonetheless, their efficacy numbers looked sensational, and it was a wonderful moment of hope for the world. No one on my team felt that, because we had not got there first, we had lost a race. Being fast was enormously important but being first was incidental. Instead, we felt a huge sense of relief. We now knew that it was possible to make a vaccine to protect people against SARS-CoV-2: something we had not known for certain the day before. A week later, on Monday 16 November, came another sensational announcement from Moderna. Their results looked if anything slightly better than Pfizer's.

At the time we were releasing a paper on safety and immune

responses in older people, which was somewhat overshadowed in the media by all the other good vaccine news. Nonetheless, what we were showing in this paper was yet another real cause for celebration. Vaccinations don't always work as well in older people. But the blood tests from our trial participants showed that even in the oldest volunteers, the vaccine was inducing good levels of the right kinds of antibodies and T cells, with no apparent differences in immune response between age groups. Because older people were so prone to serious illness and death from Covid-19, this was a crucially important finding. These were, of course, healthy older people (because this was a phase II study) and perhaps not fully representative of the general population of over-55s in the UK. But with the study showing no drop-off in immune response even in those aged over 70, at least we now knew that age itself was not a barrier to strong immune responses.

The wait for our efficacy results was finally over the following Monday, 23 November. It was a moment we had been dreaming of and waiting for. It was an enormous achievement. And yet . . . like so many other big moments that year, it was not an ending, it was just the beginning of the next phase. We were one step closer to the time when the devastation caused by SARS-CoV-2 would be brought to an end, but we were not there yet. We enjoyed the celebratory headlines. We bit our tongues at the confused or critical ones. And we tried to stay focused on our next tasks: reviewing all the data in detail again, before publishing it for scrutiny by the scientific community in the *Lancet*; and submitting it to the regulators in the UK (the MHRA) and the EU (the EMA). It would then be up to the regulators to decide whether to give the vaccine emergency-use licensure so it could be used in the real world.

There were small but important details in the data that took time to iron out. For example, a couple of severe adverse events had been coded as 'no adverse event' which clearly didn't make sense. After investigation it turned out that the events in question were both 'being admitted to hospital for planned surgery'. These should not have been recorded as severe adverse events (SAEs) so the records were corrected. This matters because the total number of SAEs must be reported, as well as what they are. As we discussed in the previous chapter, the SAEs that occur during a clinical trial usually follow the same pattern as in the general population but if there is an unexpected increase in any type of adverse event, it could be a consequence of vaccination and that then needs to be looked into.

And of course, we also needed to do more work to dig into the data and try to understand the unexpected aspect of our trial results. Was it really the first half-dose before a standard dose that resulted in higher efficacy, or something else? What was the science driving it? What were its implications?

Through the following weeks, there was in some ways a pleasing sense of symmetry that a project I had first considered on the first day of 2020 when I had read of the SARS-like pneumonia cases in Wuhan, China could be reaching a climax at the very end of December with the anticipated emergency-use licensure of the vaccine by the MHRA. But that perspective was difficult to keep hold of. On the whole, the sniping from the US biotech press, parroted by some in the UK; the increasingly anxious mood in the country as case numbers continued to rise and the logistically challenging roll-out of the Pfizer vaccine got underway; and the usual end-of-year activities were creating a rising sense of pressure within the team. We were worn out, and we all needed to plan our Covid Christmases.

The first Sunday in December was misty and cold. That morning one of my daughters had got out the sewing machine and was planning a day of sewing. She showed me a beautiful little piece of embroidery that she had done, and some fabric she had ordered, to make some small bags. I would have loved to have spent the day at home with fabric, ribbons and wool, or sorting out the Christmas decorations. But I had to be in the office for a discussion at noon about the data, and then for an online talk I was giving that evening. I felt guilty about the number of similar requests I had turned down over the past weeks.*

The following week our efficacy findings were published in the *Lancet* and my son started a complex series of manoeuvres that would, I hoped, allow him to be with us for Christmas. He had made arrangements to take the two Covid tests that students had to have before coming home from university. He also had to get to Bristol for his vaccine trial check-up, six months after his second vaccination (all three of my children had volunteered for the trial). I was delighted when he finally walked through the front door. We hugged and I admired his new haircut, which one of his flatmates had given him the day before.

My family was safely gathered but the atmosphere in the country was getting increasingly feverish. A new variant of SARS-CoV-2 then known as the Kent variant seemed to be responsible for soaring case numbers. The government was making confusing new announcements seemingly every day about different parts of the country moving into more restrictive tiers. Each tier had different and complicated rules about how

* Saying yes to every request to give a talk, or review a paper, or participate in an event, could have filled my time many times over and I had to learn to harden my heart. I tried to prioritise the talks that would reach the widest audience and have the most impact. I don't do it for the occasional gifts but after I gave a talk to the Pharmacy Guild of Australia they sent me a food hamper that was so good it led my partner Rob to ask me whether I could do any more talks for Australian pharmacists.

many people could meet indoors, in the garden, or in public spaces, and there were more rules about moving between parts of the country in different tiers. Over the weekend, the plan for five-day, three-person temporary Christmas bubbles was finally scrapped and a new tier, lockdown by any other name, was introduced. By the following Monday, with Christmas fast approaching, France had closed its borders to lorries from the UK, causing supermarkets to warn there might soon be food shortages, and the *Daily Mail* was calling for the MHRA to hurry up and approve our vaccine as it was revealed that only 500,000 people had been vaccinated since the beginning of the Pfizer roll-out nearly two weeks earlier.[1] Although there were reports of GP practices struggling with the logistics of booking in patients and then running the clinics with all the necessary social distancing, the limiting factor was not the NHS's ability to vaccinate people but the number of doses available. At this rate it would take more than a year to vaccinate just the 15 million people in the top four priority groups.

That Monday before Christmas, we made our final data submission to the MHRA. It was 18,000 pages long and the MHRA said it would take a minimum of seven days to review. We would be checking our phones all through Christmas in case of any news but I didn't expect to hear anything for a while. I also didn't think anyone at the MHRA would be getting much of a Christmas.

—

That same day I began the work to produce two new versions of our vaccine. I had hoped that we could wait until 2021 before starting all over again, but with reports of rapidly-spreading mutant strains I felt we had to start straightaway.

Over the last couple of weeks, news of variants of SARS-CoV-2 that behaved differently from the original virus had been coming in. As always, data arrived piecemeal to be put together into a full understanding, and as always, we needed to take action before we had the perfect data set. (And as always, sections of the press demanded certainty immediately.) What became clear quite quickly was that a variant then known as the Kent variant, and later renamed the Alpha variant, was more transmissible than the original virus and was rapidly moving through England. The spike protein of the virus had mutated and become better able to bind to the relevant receptor on the surface of human cells than the original virus.

This was not a particularly surprising development. It is how evolution works: random mutations, followed by the natural selection of versions with advantageous traits. A virus that has randomly mutated so that it can no longer infect people will disappear. A virus that has randomly mutated so that it infects people more easily will have an advantage and will quickly take over. Two other potentially concerning variants were also identified: the so-called South African variant, later renamed Beta, which also involved changes to the spike protein, and the so-called Brazilian, or Gamma, variant.

Although the spike protein is what we use in our vaccine, it did not look as though the new variants with slightly changed spike proteins would completely 'evade' the immune response our vaccine produced. In other words, it seemed our vaccine would still work against the new variants: if not as well as against the original virus, then well enough not to require a switch to an updated version quite yet. But, just as nearly a year earlier, in January 2020, if there was a possibility we might need it, we had to start making it, at risk and immediately.

So, we had to work out the new gene sequences to drop

into the adenovirus vector – a relatively easy and quick part of the job. We also had to work out how we could move as fast as possible at every stage: making the new starting material, conducting new clinical trials, switching over large-scale manufacturing from the original vaccine to the new one, and getting the new vaccine licensed. This all needed to have been worked through in advance so that if and when we needed to switch, we had the best and fastest possible process planned out and ready to go. I placed orders for some new DNA sequences and Cath and I started planning again.

—

My family had a quiet Christmas, which I for one was very happy with. We had a film night, cocktails before dinner on Christmas Eve, and on Christmas Day my daughters cooked an absolutely splendid Christmas lunch with all the trimmings. I watched some TV, did a bit of knitting and a bit of the Christmas jigsaw, lit the woodburning stove and snoozed on the sofa, but always keeping my phone close in case of any news. Case numbers kept rising. Hospital numbers were rising too. The papers continued to be full of impatient speculation about when our vaccine would be licensed, with predictions that it would be a 'game-changer' as there would then be so many more doses of a vaccine that was so much easier to deliver. There was obviously a tension between getting people vaccinated as soon as possible and allowing the MHRA time to complete their review. I felt it like everyone else.

On the 28th, I gave up on obeying everyone's instructions to enjoy some downtime while I could and went into the office. I knew my inbox had continued to fill up over Christmas and I was finding it difficult to switch off while we waited for news.

It was – I imagined – like waiting for the birth of a grandchild. There's nothing you can do, but you can't just forget about it and carry on as normal. My mother always complained about how stressful it was for her waiting to hear about my triplets being born. When she did, I always thought to myself that however hard it was to be waiting, giving birth was worse.

Twenty-two years earlier, I had gone into labour ten weeks early, at thirty weeks. That meant all three babies were probably going to need ventilators to help them breathe. But the hospital only had one that was suitable. There were discussions about delivering one baby and then transferring me (with the first baby on the portable ventilator?) to a different hospital with more ventilators. By this point there was no chance of transferring me before any of the babies were born – if we tried, I was warned, 'the baby will be delivered in the ambulance'. So there was a real risk of me being separated from one or more of the babies, and they might not all survive. The first baby was delivered, breathed without needing a ventilator, and was whisked off to the special-care baby unit. I spent the next sixteen hours on intravenous drugs to try to stop my labour and hooked up to multiple monitors to keep track of my contractions and the babies' heartbeats. But the high dose of the drug started to affect my heart, and when it was reduced my contractions started again. Eventually, a plan to take me to visit my firstborn was abandoned and I was instead taken back to the operating theatre to deliver the other two babies. They also breathed without assistance and were admitted to the high-dependency unit (for babies in a less critical condition than those in the special-care baby unit) where they were joined by their older sibling the next day.

This time, it was me in the role of the grandmother, waiting for the MHRA to approve our vaccine, and that was fine. I reminded myself that however hard it was to be waiting, all around the world others – healthcare workers struggling to cope with the influx of Covid patients, and patients struggling to breathe – were going through worse.

By the end of the day I knew that licensure was coming. The announcement would be made by the MHRA on the morning of Wednesday 30 December. And that is the end of the story, and we all lived happily ever after.

—

If only life could be so simple. Of course, it wasn't over. AstraZeneca had millions of doses ready but would need to make hundreds of millions more. The vaccines still had to be rolled out and put into people's arms throughout the world: a task just as huge as making the vaccine and proving it worked. And the virus was still out there – as transmission continued there was potential for more worrying mutations to arise.

You might have expected, though, that for the Oxford team, who were not overseeing global manufacturing or the vaccine roll-out, the time between the vaccine being approved by the MHRA on 30 December 2020 and the first vaccinations being administered on 4 January 2021 would have been fairly quiet. A chance to reflect on a full year of hard work, with success at the end of the year and the prospect of lots of people being vaccinated very soon. Perhaps a moment to finally congratulate ourselves on what we had accomplished. That wasn't how things turned out. Instead, just as I was metaphorically undoing my trainers with that post-run sense of achievement and pride, a wasp stung me on the ear.

We did receive some congratulations, but the media was also full of anxiety and criticism over various aspects of the vaccine roll-out: the decision to license two full doses twelve weeks apart; the speed of the roll-out; whether the vaccines would work against new variants, and so on. This was all a bit irritating but the anxiety was understandable – a record 80,000 new cases were recorded in the UK on 29 December, hospitals were close to becoming overwhelmed and it was not clear whether even another lockdown would bring things under control. And we had done our job. So the criticism wasn't too hard to ignore. That was just buzzing. Then came the real sting which could have derailed the entire vaccination programme. Potentially, roll-out of our vaccine was about to be halted before it had even begun.

As Cath explained in Chapter 5, our vaccine is manufactured using a human cell line, HEK293 cells. After we have grown the vaccine in the cells, we then separate out the bits we want to keep – the adenoviral-vectored vaccine – from the components of the cells that we don't need any more. Each batch is then tested to make sure that has happened.

However, there are other viral-vectored vaccines that are made using different types of virus that are surrounded by a lipid (fat) envelope. As those viral-vectored vaccines are also produced inside cells, during manufacturing some of the proteins from the production cells will be incorporated into the virus's lipid envelope and remain as part of the vaccine even after purification. If the production cells are of human origin, then those proteins will include human leukocyte antigens, or HLAs. HLAs are part of the immune system, and there are lots of different versions of them. We inherit our own particular combinations of them from our parents and that combination is what we are looking at when we look at someone's tissue type before they receive an organ transplant. It is the job of the immune system to distinguish between 'self', things that are part of us and are supposed to be there, and 'non-self', things that are foreign and should be attacked. If a kidney transplant takes place between identical twins, there will be no rejection of the donated kidney because identical twins have identical HLA molecules in their cells, including in the cells of the donated kidney. Between strangers, it is necessary to find as close a match as possible and the recipient will probably still need to stay on immunosuppressive drugs to prevent rejection.

If someone is vaccinated with an enveloped viral-vectored vaccine manufactured on a human cell line, then it is possible they could become sensitised to the HLA molecules from that cell line, which would make it more difficult for them to receive a transplant should they ever need to. It was pointed out to us, and to the MHRA, that this had in fact happened in a small clinical trial of a candidate vaccine against HIV. This had ended the development of that particular vaccine.

In between the MHRA giving approval for use of the vaccine and the initiation of the vaccination campaign, we suddenly found ourselves being asked to prove that this was not going to happen with our vaccine.

Of course this was a purely theoretical concern, although an important one. Adenoviruses are not enveloped viruses. They contain no lipid at all. Also, one of the final tests on each batch of vaccine is to measure the amount of host cell protein remaining, and it has to fall below a very low limit. So there was no reason to expect that intact HLA molecules might be present and might result in sensitisation. We searched the scientific literature for any publications relating to this but there were none. No one had ever reported on the absence of HLA-sensitisation after adenoviral-vectored vaccination; probably because there was no reason to expect that it might happen so no one had been motivated to find out. But, since there was nothing in the literature, that meant there was nothing reassuring in the literature.

The other approach that we could take was to measure the reactivity to HLA molecules in the blood of some of our volunteers to see if there was any sign that this could have been increased by vaccination. So, on New Year's Day 2021, staff were in the labs finding and packing up 1,200 blood samples to send off for testing. Since women who have been pregnant will become sensitised to the HLA molecules that their babies have inherited from their father, it was simplest to stick to samples from male trial participants. Six hundred samples taken at twenty-eight days after the second vaccination – when the response, if it existed, would have been strongest – were packed up, along with matching samples from the day the same volunteers received the first vaccination for comparison. It was quite a contrast from my New Year's Day of 2020, when I was relaxing at home, and browsing a few science websites to see if there was anything interesting going on. 'May you live in interesting times' is of course a curse rather than a blessing.

Amazingly, by lunchtime the next day, 400 of the pairs of samples had been analysed in an NHS lab. Everything looked normal. There was no evidence of HLA sensitisation, which was what we had expected but was now supported by data. In the following twenty-four hours the rest of the samples were analysed and the results presented to the MHRA. The theoretical-but-important issue was now definitely back in the realm of non-issue. In the meantime, I had been back in the realm of going into the office regardless of weekends or bank holidays, and keeping secrets from my family, which was something I had hoped to be putting behind me.

—

Wednesday 27 January 2021. I felt a mix of giddy elation and frustration. I hadn't quite known what to do with myself all morning. After much not-very-patient waiting, I was finally going to be vaccinated.

Throughout January there had been increasing tension in the vaccine team. There had, of course, been no happy ending yet. I was now working very hard on planning the route to making new versions of the vaccine (three of them now for the three most worrying variants), along with Cath and her team. Andy's team was still following up clinical trial participants with Covid symptoms to add to the efficacy data and taking blood samples to assess how well immunity was being maintained over time. Others in Tess's team were still in the lab every day measuring immune responses and preparing samples for shipping out to other labs for more testing. There had been no let-up, no chance to take stock or celebrate, and the long days in the lab that had felt like a fun sprint in the optimistic days of early summer had become a marathon-like slog. The healthcare workers, meaning the doctors and nurses in Andy's team who met the clinical trial volunteers face-to-face, were being vaccinated. Following a Covid outbreak at one of the sites manufacturing the UK's supplies of the Oxford AstraZeneca vaccine, people there were being vaccinated too.* This made complete sense: no one would want supplies of the vaccine to be held up because the people who were supposed to be making it were off sick or quarantining. But those who worked in Cath's CBF making new vaccines to cope with new variants, and in the immunology lab dealing with swabs from symptomatic volunteers, were not being

* This was nothing to do with handling the vaccine. Remember, we never handle SARS-CoV-2, the virus that causes Covid-19, when making the vaccine. It was just an outbreak like any other workplace outbreak, caused by someone who was infectious coming into work and transmitting the infection to their colleagues.

vaccinated. With case numbers still very high, we all felt at risk, for ourselves but also for the work we needed to do. We explored many avenues to try to get vaccines for the team, and finally an email had dropped into my inbox inviting me to book an appointment.

Joining the socially distanced queue outside the clinic, I felt elated because I was about to receive the vaccine I had started to plan just over a year ago. I had received a vaccine of my own design before, but that had been as part of a phase I safety trial. This time, the vaccine was licensed for widespread use and instead of being part of a clinical trial I would be receiving it along with everyone else. I recognised some of my colleagues behind their masks, also queuing, and we exchanged some muffled words about what a great day it was.

I felt frustrated because this was not the end. After vaccination there would be more waiting. Two weeks for immunity to kick in. And then much more waiting for transmission to drop dramatically before we could start lifting lockdown restrictions and return to anything close to normality. I filled in my form, stated and restated my NHS number, rolled up my sleeve and received my vaccine.

I felt fine going to bed that evening but at one in the morning I woke up feeling nauseous and freezing cold, despite winter pyjamas and a thick duvet. My feet were like blocks of ice. I got a glass of water and some woolly socks (handknitted by my father; thanks Dad, still looking after me), got back into bed and began to shiver violently. If I tensed my muscles I could make the shivering stop, but as soon as I relaxed to try to go back to sleep it started up again. Then I began to feel very hot. I drank more water, took some paracetamol and eventually dropped off. When the alarm went off at seven o'clock I felt normal again, though weary after a broken night. I felt grateful, though. I had experienced a few hours of discomfort, and now I was tired, but I had known what was happening (the expected reaction to the vaccine) and how long it was likely to last (a few hours). This was so much better than starting to experience Covid-19 symptoms and – as had already been the case for more than 100 million people – not knowing what lay in store.

CHAPTER 13

Disease Y: Next Time

18 February 2021–ongoing
Confirmed cases: 110.38 million–
Confirmed deaths: 2.44 million–
Vaccinations: 193.73 million–

Thursday 18 February 2021. Exactly one year after we had kicked off the 'classic plus' method to make the starting material for the first batch of our vaccine, I was once again waiting to receive a small empty-looking tube holding 100 billion strands of DNA. The DNA was the construct for the new Beta variant, and my team would once more be making it into a vaccine.

I was, as had become the norm, worried. This time we knew a lot more about how to make the vaccine. And we would be working closely with AstraZeneca from the start rather than trying to bring them on board mid-flight, which would make a big difference. Nonetheless, the pressures felt strangely similar to the last time. What was the trial design going to be? How many doses would we need? How fast could we go? Also, there were new anxieties for this year: how many more variant vaccines would we have to make? We expected to be making at least two more, but that number would only go up as more 'variants of concern' were identified.

Although it felt like an end to the pandemic could be in sight with the rapid vaccine roll-out, it also felt like missteps over new variants could set us back again. I was not sure the team could take that. Our resilience was close to crumbling.

—

There will be a next time. The next global pandemic will come. And well before the start of 2021 we were thinking about what we needed to do to be ready for it. But it also became clear in the early months of 2021, as the world started to roll out our vaccine and others, that we were not finished with SARS-CoV-2, and it was not finished with us.

There was a lot that we were still learning about this tiny pathogen that had turned the world upside down. How could we prevent spread most effectively? How could we treat the disease? Who was most susceptible? How effective were the various licensed vaccines going to be at preventing severe disease and deaths? Mild disease? Asymptomatic transmission? How long would the immunity from the vaccines last? Might one dose of two different vaccines (for example, Pfizer and AstraZeneca) be more effective than two doses of the same vaccine?

While scientists continued to learn everything we could about the original virus, and as the good news about the real-world effectiveness of our vaccine and others started to build up, we were also having to react to the unwelcome news of new variants of concern: slightly altered versions of SARS-CoV-2 that were now circulating and, in some cases, beginning to take over from the original virus.

New variants are inevitable and we had expected them, but we had not, perhaps, anticipated that they would emerge quite as fast as they did. At the CBF we had even allowed ourselves to relax a little, thinking that our job with Covid-19 was done and we could start to 'go back to normal'.

We had possibly been lulled into a false sense of security by the knowledge that coronaviruses do not have a very high mutation rate. That is, they don't make many mistakes as they make more copies of themselves, or replicate. This is because coronaviruses have the ability to 'proofread' – to detect and correct copying errors. By comparison, HIV has a high mutation rate because the HIV virus lacks proofreading capability. Influenza is the same and this is what has made it so very challenging to create effective HIV and flu vaccines. We did not have this problem with SARS-CoV-2. However, we did, at the end of 2020, have a very large amount of virus marching through the populations of the world – and particularly high levels in the UK, Brazil and South Africa. In that situation, even a statistically unlikely event is going to occur many times.

Most mutations result in faulty versions of the virus that don't work, and so no more copies can be made, and they just disappear, with no evidence they ever happened. Mutations that give a virus some kind of advantage, though, will start to take over. This might not necessarily matter if the variant that takes over is still susceptible to the immune response induced by existing vaccines. But it very much does matter if the new variants are able to evade existing vaccines.

In late 2020 and early 2021 we knew of three variants of concern. In reality there were probably more because variants were only being picked up in the countries doing a lot of viral sequencing, including the UK. The Alpha variant was more infectious than the original virus because of a change in its spike protein that allowed it to be transmitted more effectively. Because it was being transmitted more than the original virus, it quickly became common in the UK and seeded itself around the world. (In fact, what we call the original virus is probably not the original: a mutation called D614G* happened very early in the pandemic and became the major global variant. That should have been a warning.) However, by early February we had good evidence that the antibodies induced by infection with the original virus, or by vaccination with the original vaccine, still neutralised the Alpha variant. The initial CBF work to make a vaccine specifically targeting this variant was therefore deprioritised in favour of two other variants of concern: the Beta and Gamma variants.

These variants appeared to be less well neutralised by antibodies induced by existing vaccines. The mutations these variants had accumulated were stopping the antibodies generated by infection with the original virus, or by vaccination, from binding to the new version of the spike protein; a phenomenon called immune escape. It was not that vaccines would not work against the Beta and Gamma variants at all. The signs were good that the vaccines would still prevent death and severe disease. But it looked like they were unlikely to do as well in preventing mild and asymptomatic cases.

Because all the original vaccines had been made using platform technologies, we happily did not need to start making new vaccines from scratch. Instead, we could make several potential new vaccines using the methods we had used in 2020, replacing the original gene sequence for the original spike protein with gene sequences for the new mutated versions.

---

* D614G means the amino acid D, or aspartic acid to give it its full name, at position 614 in the spike protein, had been replaced by G, which is glycine.

Although some parts of the process would take just as long as they had in 2020 – we had not found a quicker way to grow production cells – other parts could be sped up considerably. With AstraZeneca fully involved throughout the process, for example, the CBF would only have to make the initial starting materials (Step 1 from Chapter 5). The manufacturing of the vaccine itself could be done by AstraZeneca, who were now hugely experienced in production of the vaccine and were capable of producing much, much larger quantities than us in almost the same amount of time. Andy would not need to be knocking on our door every five minutes asking us how many doses we were going to make for his trial, and could we please double that number.

Also, the clinical trials that we had conducted over the last year would not all need to be repeated. For our annual flu vaccine, the strains to be included are decided on every year, and the vaccine is then produced in the same way as previously. The new version goes through a small phase I/II clinical trial to demonstrate that it induces the expected immune response, at the same time as millions of doses of the vaccine are being produced. We would probably be able to follow a similar approach with our Covid vaccine. Having demonstrated the safety, immune response and efficacy of the first version, we could probably simply test the new vaccine's ability to induce antibodies that neutralised the new coronavirus variant as well as the old one, in a few hundred people. We would look at T cell responses as well. Assuming the vaccine behaved as expected, we should be able to have it ready for use as a booster shot by the autumn of 2021.

As we go forward, we will also need to test how the new variant vaccines work in people who have previously been infected with the original variant, or previously been immunised with the original vaccine. We need to make sure that when the body sees a slightly modified version of the vaccine, it recognises the differences and makes the new kinds of B and T cells against the new variant. And we will have to work out the best combination of vaccines for particular populations with particular combinations of variants circulating. We might be able to design one vaccine that works well against both the Beta and the Gamma variants (not forgetting that the existing vaccines do still work against them to some extent). We might be able to design pre-emptive vaccines against variants that we have not yet seen but that we can predict might arise and might cause problems if they did. But none of these issues feel insurmountable.

We also think it is unlikely that the virus could mutate in a way that keeps it functioning but makes our vaccine completely ineffective. That's because a change in the spike protein that is radical enough to make our vaccine completely ineffective would also, almost certainly, be so extreme as to make the virus non-functional. So although it feels like Groundhog Day, and although we are all exhausted and wondering when it will ever end, we are tackling the new variant situation with our teeth gritted and the fortifying knowledge that we know more about what we are doing and where we are headed.

We don't know how many more variants we might be asked to make over the next year, but as global vaccination programmes get going the pool of circulating SARS-CoV-2 will decline, making the chances of new variant emergence decline too. Responding to variants as they emerge will just become a routine process, part of AstraZeneca's and other vaccine manufacturers' normal product development. We will teach AstraZeneca the method of making the very first starting materials too, so that they don't have to rely on us at all. Then we can start to get back to our 'real' job – making innovative medicines for diseases that cause global problems – perhaps now with even greater urgency. Only now, we can cross Disease X off our list and move on to Disease Y.

—

Disease Y is coming. There will be a next time. It is inevitable.

Epidemiologists and specialists in zoonotic (meaning animal to human) transmission have already warned about how the trade in wildlife, and the pressures created by intensive farming on industrial scales, are creating the opportunities for Disease Y to emerge.[1] It makes sense. The more frequently humans are in close proximity with other species, especially if those animals are kept in unsanitary conditions, the more likely it is that a naturally occurring variant of an animal virus (be that chicken, pig, bat or mink), that has randomly mutated in a way that makes it able to infect a human, actually comes in contact with a human to infect. And of course, if that infected person happens to live in crowded or unsanitary conditions themselves, the chances of onward transmission are increased.

But it's easy for people like me, who get our food plastic-wrapped from Tesco, or delivered to the door in a veg box, to yell 'just ban the markets'. The fact is that this would affect the lives of real people, with kids to feed. It's one of those problems that I know I need to learn more about before criticising others and coming up with simple solutions.

We know there will be a Disease Y. But, just as Sarah and I didn't know, when we started working on Disease X in 2018, that it would be a coronavirus, we don't know what Disease Y will be. We can make some educated guesses though. It could be another coronavirus. We have seen three spill over from animals to humans in the last twenty years – SARS, MERS and

SARS-CoV-2. It could be an influenza strain with a new rearrangement of its genetic code that makes it highly transmissible or highly lethal or both. It could be a completely unknown and unstudied virus: there are an estimated 1.67 million viruses circulating in the world, and it is thought that several hundred thousand of them are capable of infecting people. Scientists have studied 263 of these: about 0.04% of pandemic threats.[2]

Equally, there is a long list of nasty pathogens that we do already know about, and that we still have no vaccines or treatments for. One of those could cause the next pandemic. In 2020, with all eyes on the Covid pandemic, other known viruses were still circulating and spilling over into humans. In 2020 there were more than a hundred outbreaks of avian flu in birds and, in China and Laos, nine cases of human infection. There were fourteen outbreaks of Crimean-Congo haemorrhagic fever across Africa, Asia and Europe; three outbreaks of Ebola in Africa; three outbreaks of MERS in the Middle East; two of Lassa fever in West Africa; and two of Nipah in Pakistan and India.[3] Nipah virus can lead to severe disease and death in humans, with a case fatality rate of 40–75%. Lassa fever, endemic in the rodent population in many parts of West Africa, is asymptomatic in 80% of infected humans but causes severe disease in the other 20% and is usually severe in late pregnancy. And in one of the 2020 outbreaks of Ebola, fifty-five of the 135 people infected died. Sarah and I have been working on projects developing adenovirus-vectored vaccines for both Lassa and Nipah, in case of a new outbreak, and with Tess we are about to start clinical trials of a new Ebola vaccine on the same platform.

Thanks to the efforts of 2020, we now know that we have the platforms on which to build safe and highly effective vaccines at great speed. And these platforms can certainly be improved further from where they are now. We can probably speed up the process for making the initial starting material for adenovirus-vectored vaccines; and it will probably be possible to make mRNA vaccines like Pfizer's and Moderna's both cheaper and more thermally stable in the future.

But it is possible – and it's tough to say, after more than a million deaths – that the next time, things could be worse. There are aspects of SARS-CoV-2 that made this pandemic particularly difficult to contain (aerosol transmission, asymptomatic transmission, and a long infectious period before the onset of symptoms being the main ones). But compared to other diseases we already know about like Nipah, Lassa and Ebola, its case fatality rate is low. It is certainly possible to envisage a future scenario in which a highly contagious and much more fatal disease appears. If it also has the transmissibility characteristics of Covid-19, it could be extremely difficult to tackle.

So, there is an enormous imperative for us to learn as much as we can from our experiences in 2020 and 2021. This pandemic was not unexpected. In my line of work, we had been expecting it, and worrying about it, for years. But it was not properly prepared for. It would be terrible to have gone through everything we have all gone through, and then find that the economic losses that have been sustained (and they will be enormous) mean that there is *still* no funding for pandemic preparedness. We need to make sure that when Disease Y arrives, we are better prepared for it than we were for X.

It seems to me that there are three broad areas that limited our response to Covid-19, and that we need to improve in order to be in a better place the next time: infrastructure (including research and manufacturing), systems (including surveillance, stockpiling and travel bans) and global cooperation and collaboration. The solutions are not necessarily cheap or easy: but nor is dealing with a pandemic. We invest heavily in armed forces and intelligence and diplomacy to defend against wars. In the same way, we need to invest in pandemic preparedness to defend against pandemics.

While we had had some funding in Oxford for research and development of our platform technologies before 2020, we had not been able to secure funds to work on speeding up the process, or to work out how to manufacture at scale. Academic vaccine projects that attract funding are almost always to make a vaccine for a specific disease, not to do the underpinning work on more general improvements, which every funding body thinks a different funding body should finance. This work could have cut months off our response time, and the amounts we were asking for (a few million pounds) look laughably tiny compared with the hundreds of billions we have had to spend on fighting this pandemic. It is also important to point out that a significant chunk of the UK funding Sarah and others at the university have received for vaccine development in recent years has come from the UK's Official Development Assistance budget – a budget that has just been cut. Of course there is always something else to spend the money on, but it should now be apparent that vaccines are a very cost-effective and powerful way to save lives. I hope it is also apparent that the belief that we in the UK do not need to worry about diseases affecting populations on the other side of the world, and that research on vaccines against these diseases only benefits other people, is wrong and dangerous.

There are also improvements we could make to the funding mechanisms that kick in once we are in a pandemic situation. In 2020, UKRI took a very successful approach to funding Covid-19 research via a rapid mechanism. It was based on existing processes and forms but cut out anything that wasn't vital. A committee of experts reviewed and approved applications, and the details of awards were made public, to maintain oversight and transparency without causing delay. Other funders remained wedded to their complex and slow application processes, which must now be overdue for a review. It is also time to look back at which gaps need to be filled, if we are to be best placed to produce a vaccine the next time we need to, and then plan to fill those gaps.

The world must improve, and spread out, its capacity to manufacture vaccines at scale. The pressure on Sandy and the AstraZeneca team to deliver a process capable of manufacturing millions of doses, starting from the existing labour-intensive small-scale lab-based work, was enormous. Finding facilities around the world able to take on the challenge of production at scale has been an extraordinary achievement, and this is certainly something that we need to improve for next time. In the UK, the government had already identified before the pandemic that its vaccine manufacturing capacity was extremely limited. As a result, the Vaccines Manufacturing Innovation Centre (VMIC) is currently being built. This facility will, once completed, be available for manufacturing of vaccines in an emergency situation and would have been a vital resource if it had been ready in time for this crisis. I am sure that the UK will not be alone in looking to develop our own robust local manufacturing capability.

Facilities are nothing without the scientists and technicians who work in them. The UK BioIndustry Association, whose help was so instrumental in getting the manufacturing project off the ground right back at the start, has identified that skills shortages are a major risk for the bioindustry sector. It is vital that we train and retain the next generation of scientists and engineers, and that means support not only for STEM subjects at university but also for training via apprenticeships and reskilling schemes. We also need to enable the basic research that allows the development of the new technologies and new understanding, by supporting our universities and having flexible science-funding schemes that allow creativity and diversity of thinking to flourish. We have proven with this vaccine project that the UK can be a global driver of innovation and we should build on that success. If the underpinning work in technology development and vaccine manufacturing has been done in advance, then the production of a new vaccine against Disease Y at speed becomes the final flourish: the cherry on top of the cake. The problem is that people are prepared to pay for the cherry, but they won't pay for the cake.

You don't have to have been working in vaccine development to have understood that the systems to respond appropriately to a global pandemic were not as good as they could or should have been. Whether it is stockpiling essential PPE for healthcare workers, or ensuring supply-chain security for vaccine production, we need better systems implemented at both national and international level before Disease Y hits us. The WHO's early warning system for disease emergence must also continue as an international effort: we must surely have learnt that no corner of the globe can consider itself safe when a new infectious disease gets a hold. I expect that there will be a global reassessment of processes to limit international travel or mitigate its effects through quarantining, since it does seem clear now that acting early on this front saves lives, and protects economies and societies from some of the impacts of prolonged, harsh lockdowns. Learning lessons from countries that successfully implemented track-and-trace systems to prevent disease spread, like South Korea, Vietnam, Taiwan and Japan, will also be crucial.[4]

We need to devise better systems to support people who need to isolate. We need to think in advance about how we will financially support those whose livelihoods are suddenly lost because of pandemic responses; and how we continue to make available public spaces and services – churches, playgrounds, school meals – rather than withdrawing them, often from very vulnerable people, just when they are most needed.

In terms of vaccines, we could streamline and pre-plan both vaccine trials and vaccine deployment. In both cases, we could pre-register volunteers. Regulatory bodies could issue pre-agreed guidance on trial design to speed up the trials approval process. Global recognition of trial design could streamline licensure and global roll-out. These things sound tedious and bureaucratic even as I write them, but the consequences of getting them right are enormous. There are vaccine stocks sitting in fridges not being used because we have failed on this front, and we know that delays cost lives.

Running through all of this is international cooperation and collaboration. As Tedros Ghebreyesus, the director general of the WHO, and many others have said, 'no one is safe until we are all safe'. If the disease is running wild in other countries then even once we have vaccinated everyone in the country we live in, we remain at risk from someone hopping onto a plane, unknowingly carrying an emerging variant that can escape existing vaccines. So, for once, the self-interested thing to do and the altruistic thing to do are the same.[5] It would be to everyone's benefit if we could distribute vaccines more evenly across the world so that the most vulnerable could be vaccinated and the pool of virus could be shrunk. The situation in India in the spring of 2021, when a huge surge in cases threatened to overwhelm health services after another highly transmissable variant of the virus emerged, has shown us there is no room for complacency. Globally, the Covax scheme is enabling richer countries to pledge financial support for vaccine purchase for poorer countries, which is a good start. It would be even better if vaccine-rich countries started giving a proportion of their vaccine stocks right now. If every vaccine-rich country provided 10% of its stocks to rapidly support vaccination in transmission hotspots, it could have a phenomenal global effect on disease reduction.

There are plenty of things we as a society can do to be better prepared for a future outbreak or pandemic, at least in terms of having vaccines ready quicker next time. But what next for Sarah and me?

Sarah will return to the vaccine projects she was working on before Covid. She has had to leave them to one side for more than a year but at least she will now be able to use what we learned in 2020 to help accelerate them. For example, might it be cost-effective, given how much flu costs the economy, to work on flu vaccine development with as much urgency as we applied to the Covid vaccine? It would require more funding upfront, and the acceptance that not everything that was tried would work, but it might be the way to make some real progress rather than continuing to limp along as we have in the past, with small projects and no joined-up approach.

I will continue making vaccines for my colleagues at the university (not just against viruses – we are also working on a project to tackle gonorrhoea, among others). And I want to try to get the capital investment to bring my facility into the twenty-first century. The dream is a new, larger facility where we can make more innovative medicines: not just vaccines but protein and gene therapies too, to tackle cancer and blindness and lung diseases, and not just for my Oxford colleagues but for academics across the UK. There are still a lot of diseases out there that need cures, and my team wants to help.

But first, we both need some time to recover from the whirlwind of the last eighteen months or so. It has been an honour and a privilege to work on this project, but it has also taken its toll. Andy, who has also kept going without pause through all of it, says it has aged him ten years. When an interviewer asked him and Sarah recently, 'And what do you look forward to doing once the pandemic is over?', they both just stared blankly at their screens, before Andy said he couldn't really remember what life used to be like. Tess said to me around the end of February 2021, 'Are we done yet'? Another colleague says he feels he has been through a traumatic experience. I know how he feels. We have all kept going because we had to, but now we are looking forward to the ebbing away of the adrenaline and cortisol that has kept us moving all year. Sarah needs to potter in her garden and dust off her running shoes. I need to buy a round of drinks, dance in a crowd, and see my daughter hug my mum. We both need to experience some days off without constant checking of emails. Then, batteries recharged, we will be ready for the next challenge.

—

We have all spoken about 'getting back to normal', but maybe we don't want to return to exactly the way things were at the start of 2020.

A pandemic of misinformation has accompanied the spread of SARS-CoV-2. How on earth did we end up in a situation in which thousands of people decided they didn't want to be vaccinated because they had received a message that the vaccine contained a microchip that could track their movements? Ironically, most of the time they received these messages on their smartphones, with location services enabled. We need to help people to become more thoughtful and discerning about where they get their information, and whose messages they trust. Scientists do not always agree with each other, but we argue our case based on data, and the papers we publish are peer-reviewed and made publicly available. We are prepared to back up our claims with evidence, and to change our minds if the evidence changes. If you are thinking of buying a new pair of leggings, the latest teen millionaire on TikTok or a former actor selling smelly candles may be an excellent source of advice. If you are concerned about your health, you need to hear from someone who can back up what they are telling you with solid data.

More positively, the pandemic has also driven an interest in and respect for science and scientists that I hope will endure. I would love to think that, once the immediate crisis has subsided, we will continue to see scientists held up as inspirational role models, and photographed in fancy clothes for our lists of women (and men) of the year.

And, although there is still a long way to go with communicating science to the general public, my impression is that people now want to know more. The media has a role to play here and so do scientists themselves. We have to do better in providing information that is well explained without being dumbed down. It does, though, also come back around to money. While scientists are struggling with short-term contracts, and under pressure to publish research quickly in order to keep their jobs, ticking the public engagement box is rarely going to be their highest priority. Only when scientific careers are better supported and less precarious can we expect scientists to do more to help others understand what we do.

In the wider world, many people who worked in offices in cities will not return to exactly the same way they lived and worked before the pandemic. Some will look for alternatives to the time-consuming and expensive commute, the Pret lunch and £3 latte and the frequent updates to their work wardrobe. As we have learned that we can live without frequent flying and fast fashion, we have learned too how much we value our schools and hospitals. Both will need extra resources to recover from the damage and neglect incurred during the pandemic. Perhaps as societies begin to open up again, some of that support will come from people who have wanted to volunteer to do something positive during the pandemic, whether helping at vaccination centres, or supporting those in need in their local community. People want their communities to recover and are prepared to take action to achieve that. We should look for ways to help people to contribute.

Since the pandemic began, we have been reminded, forcefully and painfully, on a daily basis, that a microscopic virus spilling over from one species to another in one corner of the planet can have a catastrophic effect on the whole world. Equally, from one university campus on the outskirts of a small city, in one year, a very safe, highly effective, low-cost vaccine now saving lives around the world was created, tested and licensed for use. However we choose to impose national boundaries, and to join or leave international organisations, we cannot ignore the fact that we all share one planet. As we face into the future, and turn our attention back to climate change, poverty, war, and other intractable problems that never went away, the pandemic should act as a timely reminder of the global nature of both our biggest challenges and our most powerful solutions.

Acknowledgements

As we hope we have made clear throughout this book, the work done to create, develop, manufacture and obtain emergency-use approval for the Oxford AstraZeneca vaccine was that of a huge and diverse team, and there are many people that we must thank.

Thanks from Sarah

Firstly, my personal thanks to Andrew Pollard, the Chief Investigator of the Oxford-led clinical trials, who has compressed ten years of work into one, and never failed to do everything he could to make the vaccine available to start saving lives as soon as possible. Also to Tess Lambe, my constant ally as we battled against everything that was thrown at us, and Pedro Folegatti, stalwart Clinical Research Fellow who met all of the challenges in 2020 and afterwards with great fortitude. Thanks also to my family who supported, fed and pampered me; Rob, our three amazing children, and Sue; and to Kate for getting me out into the countryside for some respite.

Thanks from Cath

I would also like to thank Tess, who helped me stay mostly good humoured through some tough weeks, and Sandy, whose determination and drive continues to push the CBF team to aim higher and achieve more. I continue to learn enormously from the CBF team: from Richard and Eleanor about all aspects of compliance and quality; from Emma and Cathy to be calm and organised and efficient; from Omar to be business-like and ahead of deadlines; and from the rest of the team to remain cheerful in adversity and to just get on with it even when things are not quite ideal. I must also mention my research team at the Wellcome Centre for Human Genetics: Daniela, Maciej, Maria, Julia and Paulina, who put up with the fact that I deserted them for a large part of 2020, and who kept the work there ticking over nicely without much input from me. Also thanks to Leo for the line about fame, and for looking at legal things for me whenever I've asked; Bernie for some (virtual) normality; Flo and the crew at my favourite pub (@magdalen_arms) for takeaway negroni and all the pies and pasta; and the staff at Ellie's school for key worker provision when I really needed it.

On top of a pandemic, I had personal struggles this year. The Banging Lockdown Clams (love you ladies) and the Real Oxford Crewp kept me fed and sane: to Sally, Mark and Lili, Ola, Rob, Stephanie, Matthew, Emma and Ellie, Gabi and Ruairidh, Cath and Pete, Kirsty and Gary, Hayley and Evan, my gratitude and my thanks. That we Zoomed and walked and drank and laughed and sometimes cried, kept my head above water and my focus on reality when I ever tended to despair. My Zoom baking family: my sister Frances, and Paul, Hannah and Danny, and Mum and Dad – while the calories were not always wise, the advice always was, and though the bakes sometimes failed the laughter never did. It's strange that being apart made us closer, but I am glad of it (and can't wait to do last Christmas with you all soon). And finally, my daughter Ellie has been my rock this year. She's a wonderful person to just hang out with – luckily, as a lot of the year it was just the two of us. Ellie: thank you for looking after me with coffee and hugs, for doing so well at home school, and for letting me spend my weekends writing this book.

Thanks from us both

For the production of this book, we are enormously grateful to Deborah Crewe, our wonderful writer/editor, who turned our musings and jottings into a book. Thanks to Neil Blair and Rory Scarfe at the Blair Partnership who encouraged us to write the book and showed us how to make it possible. And thanks to Anna Baty, our editor at Hodder & Stoughton who proved that book production can also be speeded up. Also at Hodder, thanks to Eleni Lawrence, Vickie Boff and Claudette Morris. And to freelancers David Milner, Toby Clark, Colin Hynson, Jonathon Price and Louise Radok.

At Oxford, we thank the whole CBF team: Ioana Baleanu, Alexander Batten, Ema Begum, Eleanor Berrie, Emma Bolam, Elena Boland, Tanja Brenner, Brad Damratoski, Chandrabali Datta, Omar El Muhanna, Richard Fisher, Pablo Galian-Rubio, Gina Hodges, Frederic Jackson, Shuchang Liu, Lisa Loew, Gretchen Meddaugh, Róisín Morgans, Victoria Olchawski, Cathy Oliveira, Helena Parracho, Emilia Reyes Pabon, Gary Strickland, Abdessamad Tahiri-Alaoui, Richard Tarrant, Keja Taylor, Oto Velicka, Paul Williams, Dalila Zizi, and Sue Morris, who stepped in when we needed her.

The Jenner team: Jeremy Aboagye, Elizabeth Allen, Jordan Barrett, Sandra Belij-Rammerstorfer, Duncan Bellamy, Adam Berg, Cameron Bissett, Mustapha Bittaye, Nicola Borthwick, Amy Boyd, Federica Cappuccini, Wendy Crocker, Mehreen Datoo, Sophie Davies, Nick Edwards, Sean Elias, Katie Ewer, Sofiya Fedosyuk, Amy Flaxman, Julie Furze, Michelle Fuskova,

Ciaran Gilbride, Leila Godfrey, Giacomo Gorini, Gaurav Gupta, Stephanie Harris, Susanne Hodgson, Mimi Hou, Alka Ishwarbhai, Susan Jackson, Carina Joe, Reshma Kailath, Baktash Khozoee, Colin Larkworthy, Alison Lawrie, Yuanyuan Li, Amelia Lias, Raquel Lopez Ramon, Meera Madhavan, Rebecca Makinson, Emma Marlow, Julia Marshall, Angela Minassian, Jolynne Mokaya, Hazel Morrison, Richard Morter, Nathifa Moya, Ekta Mukhopadhyay, Andrés Noé, Fay Nugent, Marco Polo Peralta Alvarez, Ian Poulton, Claire Powers, David Pulido-Gomez, Fernando Ramos Lopez, Thomas Rawlinson, Adam Ritchie, Louisa Rose, Indra Rudiansyah, Ahmed Salman, Stephannie Salvador, Helen Sanders, Iman Satti, Jack Saunders, Rameswara Segireddy, Hannah Sharpe, Emma Sheehan, Sarah Silk, Holly Smith, Alexandra Spencer, Lisa Stockdale, Rachel Tanner, Iona Taylor, Yrene Themistocleous, Nguyen Tran, Adam Truby, Aadil El-Turabi, Marta Ulaszewska, Marion Watson, Danielle Woods, Andrew Worth, Daniel Wright, Marzena Wroblewska.

Also the team at the Oxford Vaccine Group. We chiefly worked with Maheshi Ramasamy, Parvinder Aley and Sagida Bibi, but the whole team did an incredible job.

Other important scientific collaborators were Vincent Munster and Neeltje van Doremalen at Rocky Mountain Labs, National Institutes of Health in the United States, Stefania Di Marco and her team at Advent in Italy, Sue Ann Costa Clemens in Brazil, Shabir Mahdi in South Africa, David Matthews in Bristol, Max Crispin in Southampton and other colleagues in Oxford: Brian Angus, Richard Cornall, Adrian Hill, Richard Liwicki, Sally Pelling-Deeves, Gavin Screaton, Dave Stuart. Also many thanks to Martina Micheletti for carrying more of the VaxHub work than anyone had anticipated, and to Netty England and all her colleagues at the BIA for their crucial support.

We are very grateful for the help of the press office and Public Affairs Directorate in Oxford, especially James Colman, Alex Buxton and Steve Pritchard, who helped us with all the public engagement that was so important to get right, and the fantastic Science Media Centre in London. Also the Development Office team led by Carly Nieri, who coordinated philanthropic support for our work. And of course many thanks to all of the donors, for providing the funding to fill all the gaps.

Finally our thanks go to all those who volunteered to take part in or help out with clinical trials of our vaccine, and all the other vaccines; and to AstraZeneca for stepping up to bring this vaccine to the world, partnering with us, forgoing profits during the pandemic and making a commitment to supply the vaccine to poorer countries without profit even after the pandemic ends.

Appendix A

Different types of vaccine

Vaccines are made in many different ways but ultimately all work on the same principle. They give the immune system a memory of what a dangerous pathogen looks like, so that it can tackle that pathogen if it encounters it in the future. They do this by presenting a harmless mimic of the pathogen, or sometimes just a key part of the pathogen, to your immune system.

In the case of Covid vaccines, that key part of the pathogen, that every vaccine is showing to your immune system, is the spike protein.

CONVENTIONAL VACCINES

The following are examples of conventional vaccine technologies.

Live attenuated

This type of vaccine uses a whole, live pathogen that can spread through the body after vaccination, but which has been weakened so that it does not cause disease in the vast majority of people. The immune system controls the infection caused by the vaccine, and in doing so an immune memory is formed.

When the pathogen itself is encountered, the immune response is then quickly reactivated to control the infection.

Examples currently in use are the oral polio vaccine, the Bacillus Calmette–Guérin (BCG) vaccine against tuberculosis and the FluMist nasal vaccine against influenza which is used in children.

These vaccines are not recommended for people with very weak immune systems. We used to refer to them simply as live attenuated vaccines but now, to differentiate them from some of the newer technologies, we can call these live attenuated *replication-competent* vaccines.

Inactivated

These vaccines, which use inactivated pathogens, don't cause infections. However, the immune system responds to the inactivated pathogen and makes immune responses that can then act against the live pathogen.

Most of the influenza vaccines used across the world are inactivated vaccines, made by producing large amounts of influenza virus and then chemically inactivating it. Other examples currently in use are the polio vaccine that is given by injection and the rabies vaccine.

Subunit/recombinant protein/virus-like particle

These vaccines are not made from the whole pathogen, but use one or more proteins found in that pathogen which are synthetically produced.

The hepatitis B vaccine was the first vaccine of this type to be licensed, and consists of a single viral protein which is produced in yeast cells and assembled into small particles that resemble a virus (hence virus-like particle vaccine). The human papillomavirus (HPV) vaccine is also a virus-like particle vaccine.

There are influenza vaccines consisting of recombinant haemagglutinin – haemagglutinin being one of the proteins found on the surface of the influenza virus. The Novavax Covid-19 vaccine is a recombinant virus-like particle vaccine. These types of vaccine are often used with an adjuvant.

Adjuvant

An adjuvant is not a type of vaccine. It refers to something that is added to a vaccine to increase the immune response. The most widely used type of adjuvant is alum, which is present in hepatitis A and B vaccines. Another adjuvant, MF59, is an oil-in-water emulsion and is used in an influenza vaccine sometimes given to older adults. The Novavax Covid-19 vaccine uses a different adjuvant known as Matrix-M1. This contains saponin, which is extracted from tree bark, plus cholesterol and phospholipid formed into tiny particles.

Toxoid

Some bacterial pathogens produce toxins (poisonous proteins) which then cause disease. To protect against the disease, we need to make an immune response against the toxin rather than against the pathogen itself. Toxoids are inactivated toxins: they have been modified to resemble the toxin without causing illness. Diphtheria and tetanus vaccines are toxoid vaccines.

Conjugate

In biological contexts, 'conjugate' means connected or joined. An immune response against the polysaccharides (complex sugar molecules) which are found on the surface of some bacteria can be protective. However, using the polysaccharides alone does not produce a strong immune memory. When the polysaccharides are joined onto (conjugated to) a carrier protein, though, the immune response to vaccination is stronger and an immune memory is formed. The pneumococcal vaccine is a conjugate vaccine.

PLATFORM TECHNOLOGIES

This means a technology that can be used to make vaccines against many different diseases. Once the platform is well understood, the development of a new vaccine can proceed rapidly, because much of the work has already been done. The original pathogen is not used in the production of these vaccines. We identify the antigen (the part of the pathogen against which we want to make an immune response). We then produce synthetic DNA that provides the instructions to make the antigen, and add it to the platform technology to make the specific vaccine.

The following are all examples of platform technologies.

DNA

DNA vaccines consist of loops of DNA containing the instructions to produce an antigen. The DNA is injected into muscle, where some cells will take up the DNA, make the protein and initiate the immune response. Although they have been tested in many clinical trials there are no DNA vaccines licensed for use in humans, as the immune response that is stimulated is not very strong.

RNA or mRNA

Our genes are made of DNA. Inside the cells of our body, DNA is first copied into RNA before the instructions to make a specific protein can be carried out. This type of RNA is known as mRNA, with the 'm' standing for 'messenger'. When an RNA vaccine gets inside a cell after vaccination, it instructs the cell to make a specific protein in the same way. Unlike DNA, RNA is a very unstable molecule, and so RNA vaccines are encased in fatty droplets to stabilise them and help the RNA get inside a cell after vaccination. The RNA naturally degrades and is expelled by the body after a few days. The Covid vaccines made by Pfizer and Moderna are both RNA vaccines.

Replication-deficient adenoviral vector

This is made from an adenovirus, which would normally cause a mild respiratory or gastrointestinal illness. At least one of the adenovirus genes has been removed, so that the vaccine cannot replicate and spread through the body. That is, it is replication-deficient. Then the gene for the vaccine antigen is added to the adenovirus. The antigen is then produced inside the body after vaccination.

There are many different adenoviruses that infect humans, and in people who have been infected and made an immune response to the adenovirus, the vaccine does not work quite so well. To avoid that problem vaccine developers can use either a rare human adenovirus, or an adenovirus that does not normally infect humans. ChAdOx1 nCoV-19 is based on an adenovirus found in chimpanzees. Johnson & Johnson use human adenovirus 26 in their Covid vaccine, CanSino use human adenovirus 5, and the Sputnik vaccine from the Gamelaya Institute uses both Ad5 and Ad26.

Replication-deficient poxviral vector

The Vaccinia virus which was used to eradicate smallpox is replication-competent, but there are several modified versions which are replication-deficient and can be used to produce proteins from other pathogens. This type of vaccine has been tested in many clinical trials, and Johnson & Johnson's licensed

Ebola vaccine consists of Ad26 and then a second vaccination eight weeks later with a replication-deficient poxviral vector, each of which produces the Ebola glycoprotein after vaccination.

Replication-competent viral vector

The measles vaccine is a live attenuated replication-competent vaccine, which can also be used to carry genes from other pathogens, although there are no licensed vaccines made in this way at present. Merck's licensed Ebola vaccine uses another replication-competent virus, vesicular stomatitis virus (VSV). Unlike other platform technologies, the VSV-based vaccine carries the Ebola glycoprotein on the outside of the VSV virus, rather than carrying the gene into the body so that the protein is produced inside the body.

Appendix B

The classic method and the rapid method

THE CLASSIC METHOD

1. Start in the research lab. Take the DNA for ChAdOx1 – your vector – and, using genetic engineering techniques, **recombine** it with the DNA for the part of the virus you want to vaccinate against. If you are making a vaccine against Covid-19, this would be the DNA that codes for the spike protein. Now you have circular pieces of DNA, also known as plasmids or bacterial artificial chromosomes, containing the whole genetic sequence of the vaccine you want to make.

2. In order to create larger quantities of your plasmids, **insert** them into specially treated bacteria. This process is known as transformation. As the bacteria divide and grow (in their nutritious broth), they will make more copies of the plasmid DNA.

3. **Purify** (meaning remove the constituent parts of the bacterial cell like proteins, membranes and bacterial chromosomes) so that you are left with pure plasmid DNA. The next steps need to be done in very controlled conditions, so at this point, ship the DNA to the CBF.

4. The purified DNA contains some sequences that we don't need in the final vaccine: the genes that enable it to copy itself inside bacteria. Using restriction enzymes (tiny chemical scissors), make two precise **cuts** around the unwanted piece of DNA. Discard it.

5. Your remaining linear piece of DNA, the ChAdOx1-vectored vaccine DNA sequence, is the blueprint for your vaccine. However, you only have tiny amounts of it. Virus particles cannot assemble or replicate outside living cells, so you need to **insert** your viral DNA sequence into living cells, which will act as little microfactories to make the vaccine. This process is known as transfection. We use specially cultured human cells called HEK293 cells for this. These cells contain the adenovirus gene E1. This gene allows the ChAdOx1 adenovirus – which has had its E1 gene removed so that it cannot replicate itself inside normal human cells and cause an infection when used as a vaccine – to replicate itself inside these cells.

6. To do this, **mix** your DNA with a solution to allow it to get inside the HEK293 cells. This is not a very efficient process. However, as long as you get enough of your DNA into some HEK293 cells, that DNA will then instruct those cells to make lots of copies of the adenovirus, which is then very efficient at spreading itself to infect more cells.

7. **Incubate** your mixture (you should have about a million HEK293 cells in 2 ml of growth medium) at 37 degrees (i.e. body temperature) for about a week. During this time, more and more cells become infected and make more and more adenovirus particles.

8. **Amplify** the virus culture by transferring to new cells in a larger container to make even more adenovirus particles.

9. **Release** the virus from the cells into a solution and measure the concentration of infectious viral particles.

10. **Dilute** the virus solution and use it to **infect** lots of tiny cell cultures. Aim to infect each culture with just a single virus particle, so that the resulting vaccine preparations are each derived from a single independent virus – i.e. are individual clones. (This process is called single-virion cloning.)

11. **Incubate** each culture again.

12. **Amplify** again.

13. **Purify** the virus (remove the complicated mixture of proteins, nucleic acids, fats and carbohydrates that make up the contents of a human cell) so that you are left with pure viral vaccine particles.

14. **Test** your purified viral particles carefully to find one with a DNA sequence that is completely homogeneous and correct. (There are always some errors in the synthetic DNA provided, and also, mutations can arise during the manipulations you have done to get this far. So we have learnt that to give ourselves the best chance of having at least one correct vaccine at the end, we should run steps 10 to 13 in parallel multiple times. Even so, we have sometimes had to go back and repeat these steps two or three times to get one correct vaccine clone.)

15. **Expand** the correct clone in larger and larger volumes of culture.

16. What you have now is precious material. It is the pre-GMP starting material that will go on to seed all the manufacturing of this vaccine. **Check** it to ensure it is correct, sterile and free from contaminants then **certify** it before **transferring** to a GMP-manufacturing facility.

THE RAPID METHOD

1. **Use the classic method** to make, in advance, a ChAdOx1-vectored vaccine that contains green fluorescent protein (GFP) instead of a gene from a virus. (If you have ever seen pictures of glowing green rabbits or fish, it is because they have been genetically engineered to produce GFP, which glows green under light of the right wavelength. It is useful here because, if our ChAdOx1 vaccine produces GFP, then we can see which cells it has infected by looking down an ultraviolet microscope and seeing which cells glow green.)

2. Now you have a small stock of a vaccine that is 100% genetically correct and of a suitable quality to vaccinate someone with. We aren't going to do that though because we don't want to make people glow green.

3. Instead, **purify** the DNA from inside the adenovirus.

4. Use your restriction enzymes (chemical scissors) to **cut** out the GFP gene.

5. **Test** your stock again carefully. Make sure there are no contaminants in it. There is a theoretical risk that the GFP gene did not get cut out completely: test by using a small amount to infect some cells and then checking them under a fluorescence microscope. None of them should glow green.

6. **Set aside** until you are ready to use.

7. As soon as you receive a newly synthesised gene for a new vaccine, you can **add** it into the DNA from step 6 that you prepared earlier and allow an assembly reaction to occur.

8. You now have your vaccine blueprint, within hours of receiving the new gene. This can be directly introduced into the human cells that are needed to allow virus replication and assembly. Simply transfect: **mix** it with a solution to allow it to get inside HEK293 cells. The process will be quicker and more reliable than in the classic method because adenoviral DNA that has been purified from a virus, rather than from a copy of the viral DNA on a plasmid made in bacteria, works much more efficiently to generate the new virus once inside the cell.

9. Because the process is so much more reliable, you can set up your transfection (i.e. the process of getting the virus inside the human cells) so that only one cell at a time will be transfected. This means that you can **skip steps 10 to 12 of the classic method**, which are designed to achieve the same thing: starting material derived from a single cell, transfected with a single virus particle.

10. You now have lots of different vaccine stocks, each derived from one transfected cell. It is likely that the synthetic DNA contained some incorrect strands. Don't worry. **Expand** your independent, clonal vaccine stocks, **test** them, and choose one that is 100% genetically correct.

11. Continue to **expand** your favourite clone in larger and larger volumes of culture.

12. You now have the precious pre-GMP starting material that will go on to seed all the manufacturing of this vaccine. As in the classic method, **check** it to ensure it is correct, sterile and free from contaminants, **certify** it, and **transfer** it to a GMP-manufacturing facility.

Appendix C

What is in the Oxford AstraZeneca vaccine?

This is the information that appears in the patient information leaflet for the Oxford AstraZeneca vaccine, annotated by us to make it easier to understand. Anything that is not listed is not in it.

> At 0.5 ml per dose, you could fit ten doses on a teaspoon.

One dose (0.5 ml) contains: Covid-19 Vaccine (ChAdOx1-S* recombinant) 5×10^{10} viral particles

> This means each dose contains 50,000,000,000 viral particles.

* Recombinant, replication-deficient chimpanzee adenovirus vector encoding the SARS-CoV-2 Spike glycoprotein.

> See pages 38–39 and pages 64–70 for explanations of each of these terms. In this book, for simplicity, we refer to the spike protein rather than the spike glycoprotein. However, strictly speaking, the SARS-CoV-2 spike protein is actually a glycoprotein, that is, a protein with carbohydrates (sugars) attached.

Produced in genetically modified human embryonic kidney (HEK) 293 cells.

> The reason we produce the vaccine in HEK cells is set out on page 101. The Vatican's attitude towards this is described in Chapter 8.

> The genetically modified organisms are the viral particles. The process of genetically modifying them is described on pages 38–39 and in Chapter 3 and Appendix B.

This product contains genetically modified organisms (GMOs).

The other excipients are:

> Excipients means ingredients other than the active ingredients. In this case the active ingredients are the viral particles.

- L-histidine

> This is an amino acid found in almost every protein in the human body. It is used during manufacture to keep the pH balance of the vaccine right.

- L-histidine hydrochloride monohydrate

> This is a different form of the histidine, also used to keep the pH balance of the vaccine right.

- magnesium chloride hexahydrate

This is a salt. It is used as a stabiliser for the DNA in the vaccine.

- polysorbate 80

This is a detergent. It is used as a stabiliser for the viral particles in the vaccine.

- ethanol

This is alcohol. It is used as a solvent and present in tiny amounts: 0.002 mg per dose. The British Islamic Medical Association statement on the vaccine says, 'This is "not enough to cause any noticeable effects" and has been described as negligible by Muslim scholars. It is comparable or less than the amount of ethanol found in natural foods or bread, for example.' See https://britishima.org/covid19-vaccine-az/.

- sucrose

Sugar. It is used as a stabiliser.

- sodium chloride

Salt – like the salt we put on our food. It is used to stabilise the viral particles. The various stabilising ingredients (the salts and the sugar) make the solution more like the inside of a human cell, which is where the virus is happiest.

- disodium edetate dihydrate

This is used as a preservative. It prevents anything unwanted from growing in the solution, and prevents any enzymes from degrading the adenovirus. It is also used in eyedrops and in some foods.

- water for injections.

Extremely pure water.

Notes

Covid statistics

Except where indicated otherwise, the source for confirmed Covid cases, confirmed deaths from Covid, and vaccinations given throughout the book is the Our World In Data website, as at 26 April 2021.

https://ourworldindata.org/grapher/cumulative-deaths-and-cases-covid-19?country=~OWID_WRL

https://ourworldindata.org/covid-vaccinations?country=~OWID_WRL

Published papers on the Oxford AstraZeneca vaccine

More detail about the science behind the Oxford AstraZeneca vaccine, ChAdOx1 nCoV-19, also known as AZD1222, can be found in the published papers listed below.

Lancet papers on clinical safety, immunogenicity and efficacy

Folegatti PM, Ewer KJ, Aley PK, et al. 'Safety and immunogenicity of the ChAdOx1 nCoV-19 vaccine against SARS-CoV-2: a preliminary report of a phase 1/2, single-blind, randomised controlled trial', *Lancet* 2020; 396(10249): 467–78.

Ramasamy MN, Minassian AM, Ewer KJ, et al. 'Safety and immunogenicity of ChAdOx1 nCoV-19 vaccine administered in a prime-boost regimen in young and old adults (COV002): a single-blind, randomised, controlled, phase 2/3 trial', *Lancet* 2021; 396(10267): 1979–93.

Voysey M, Costa Clemens SA, Madhi SA, et al. 'Safety and efficacy of the ChAdOx1 nCoV-19 vaccine (AZD1222) against SARS-CoV-2: an interim analysis of four randomised controlled trials in Brazil, South Africa, and the UK', *Lancet* 2021; 397(10269): 99–111.

Voysey M, Costa Clemens SA, Madhi SA, et al. 'Single-dose administration and the influence of the timing of the booster dose on immunogenicity and efficacy of ChAdOx1 nCoV-19 (AZD1222) vaccine: a pooled analysis of four randomised trials', *Lancet* 2021; 397(10277): 881–91.

Emary KRW, Golubchik T, Aley PK, et al. 'Efficacy of ChAdOx1 nCoV-19 (AZD1222) vaccine against SARS-CoV-2 variant of concern 202012/01 (B.1.1.7): an exploratory analysis of a randomised controlled trial', *Lancet* 2021; 397(10282): 1351–62.

Clinical immunology papers

Barrett JR, Belij-Rammerstorfer S, Dold C, et al. 'Phase 1/2 trial of SARS-CoV-2 vaccine ChAdOx1 nCoV-19 with a booster dose induces multifunctional antibody responses', *Nat Med* 2021; 27(2): 279–88.

Ewer KJ, Barrett JR, Belij-Rammerstorfer S, et al. 'T cell and antibody responses induced by a single dose of ChAdOx1 nCoV-19 (AZD1222) vaccine in a phase 1/2 clinical trial', *Nat Med* 2021; 27(2): 270–8.

Madhi SA, Baillie V, Cutland CL, et al. 'Efficacy of the ChAdOx1 nCoV-19 Covid-19 Vaccine against the B.1.351 Variant', *N Engl J Med* 2021.

Zhou D, Dejnirattisai W, Supasa P, et al. 'Evidence of escape of SARS-CoV-2 variant B.1.351 from natural and vaccine-induced sera', *Cell* 2021.

Supasa P, Zhou D, Dejnirattisai W, et al. 'Reduced neutralization of SARS-CoV-2 B.1.1.7 variant by convalescent and vaccine sera', *Cell* 2021; 184(8): 2201-11 e7.

Dejnirattisai W, Zhou D, Supasa P, et al. 'Antibody evasion by the P.1 strain of SARS-CoV-2', *Cell* 2021.

Preclinical papers

Van Doremalen N, Lambe T, Spencer A, et al. 'ChAdOx1 nCoV-19 vaccine prevents SARS-CoV-2 pneumonia in rhesus macaques', *Nature* 2020; 586(7830): 578–82.

Silva-Cayetano A, Foster WS, Innocentin S, et al. 'A booster dose enhances immunogenicity of the COVID-19 vaccine candidate ChAdOx1 nCoV-19 in aged mice', *Med (N Y)* 2021; 2(3): 243–62 e8.

Van Doremalen N, Purushotham J, Schulz J, et al. 'Intranasal ChAdOx1 nCoV-19/AZD1222 vaccination reduces shedding of SARS-CoV-2 D614G in rhesus macaques', *bioRxiv* 2021.

Graham SP, McLean RK, Spencer AJ, et al. 'Evaluation of the

immunogenicity of prime-boost vaccination with the replication-deficient viral vectored COVID-19 vaccine candidate ChAdOx1 nCoV-19', *NPJ Vaccines* 2020; 5(1): 69.

Almuqrin A, Davidson AD, Williamson MK, et al. 'SARS-CoV-2 vaccine ChAdOx1 nCoV-19 infection of human cell lines reveals low levels of viral backbone gene transcription alongside very high levels of SARS-CoV-2 S glycoprotein gene transcription', *Genome Med* 2021; 13(1): 43.

Fischer RJ, van Doremalen N, Adney DR, et al. 'ChAdOx1 nCoV-19 (AZD1222) protects hamsters against SARS-CoV-2 B.1.351 and B.1.1.7 disease', *bioRxiv* 2021.

Watanabe Y, Mendonça L, Allen ER, et al. 'Native-like SARS-CoV-2 spike glycoprotein expressed by ChAdOx1 nCoV-19/AZD1222 vaccine', *ACS Cent. Sci.* 2021, 7, 4, 594–602.

More information

For more detail on the effectiveness of the vaccines against SARS-CoV-2 see:

https://www.gov.uk/government/publications/phe-monitoring-of-the-effectiveness-of-covid-19-vaccination

https://publichealthscotland.scot/news/2021/february/vaccine-linked-to-reduction-in-risk-of-covid-19-admissions-to-hospitals/

For more detail about safety reporting through the Yellow Card scheme in the UK see:

https://www.gov.uk/government/publications/coronavirus-covid-19-vaccine-adverse-reactions

For a wide range of general information on Covid vaccines see:

https://www.who.int/news-room/q-a-detail/coronavirus-disease-(covid-19)-vaccines

http://vk.ovg.ox.ac.uk

Chapter 1

1. *The Times*: 'This is a remarkable achievement for British science and offers hope to the world of an end to the pandemic', https://www.thetimes.co.uk/article/the-times-view-on-the-success-of-the-oxford-vaccine-great-british-breakthrough-ljwltmtbc. *Guardian*: 'Vaccine results brings us a step closer to ending Covid', https://www.theguardian.com/world/2020/nov/23/vaccine-brings-us-a-step-closer-to-ending-covid-says-oxford-scientist. *Financial Times*: 'Vaccine cements Oxford place as leader in battle against Covid', https://www.ft.com/content/f147199b-11e6-444b-9514-94352bded128. *Daily Express*: 'Jubilation at Oxford vaccine breakthrough', https://www.pressreader.com/uk/daily-express/20201124/281496458834440. *Daily Mirror*: 'Harsh winter . . . brighter spring', https://www.pressreader.com/uk/daily-mirror/20201124/281496458834443. *Daily Mail*: 'Vaccine cheers . . . but first the tiers. Oxford jab is triumph for UK', https://www.dailymail.co.uk/news/article-8979771/Harsh-Covid-restrictions-remain-April-despite-stunning-vaccine-breakthrough.html. *Metro*: 'Get yourself a vaccaccino: The Oxford jab will cost less than a cup of coffee', https://www.pressreader.com/uk/metro-uk/20201124/282505776149067.
2. *New York Times*: 'After admitting mistake, AstraZeneca faces difficult questions about its vaccine', https://www.nytimes.com/2020/11/25/business/coronavirus-vaccine-astrazeneca-oxford.html. *Wired*: 'The AstraZeneca vaccine data isn't up to snuff', https://www.wired.com/story/the-astrazeneca-covid-vaccine-data-isnt-up-to-snuff/.
3. https://www.statnews.com/2020/11/23/astrazeneca-covid-19-vaccine-is-70-effective-on-average-early-data-show/.
4. *The Times*: 'AstraZeneca defends Oxford vaccine as disquiet mounts over the results', https://www.thetimes.co.uk/article/astrazeneca-defends-oxford-coronavirus-vaccine-as-disquiet-mounts-over-the-results-mf6t57rnr. *Daily Telegraph*: 'Manufacturing error clouds Oxford's Covid-19 vaccine study', https://www.telegraph.co.uk/news/2020/11/26/manufacturing-error-clouds-oxfords-covid-19-vaccine-study-results/. *Financial Times*: 'Doubts raised over Oxford-AstraZeneca vaccine data', https://www.ft.com/content/4583fbf8-b47c-4e78-8253-22efcfa4903a.
5. https://www.barrons.com/articles/sanofi-and-glaxo-delay-their-covid-vaccine-pfizer-and-moderna-extend-their-leads-51607701067.
6. https://ourworldindata.org/covid-vaccinations?country=~GBR.

7. https://www.theguardian.com/society/2020/dec/18/nhs-staff-priority-covid-vaccine-hospital-bosses-england-coronavirus.
8. Jonny Dimond, *World At One*, 30 December 2020; Mishal Husain, *Today* programme, 30 December 2020; Jonny Dimond, *World At One*, 30 December 2020.
9. https://www.handelsblatt.com/politik/deutschland/pandemie-bekaempfung-corona-impfstoff-diskussion-um-wirksamkeit-von-astra-zeneca-vakzin-bei-senioren/26849788.html
10. https://www.politico.eu/article/coronavirus-vaccine-europe-astrazeneca-macron-quasi-ineffective-older-pe/.
11. https://ir.novavax.com/news-releases/news-release-details/novavax-covid-19-vaccine-demonstrates-893-efficacy-uk-phase-3 and https://www.nih.gov/news-events/news-releases/janssen-investigational-covid-19-vaccine-interim-analysis-phase-3-clinical-data-released.
12. https://papers.ssrn.com/sol3/papers.cfm?abstract_id=3789264 and https://publichealthscotland.scot/news/2021/february/vaccine-linked-to-reduction-in-risk-of-covid-19-admissions-to-hospitals/.

Chapter 2

1. Unless otherwise stated, case numbers in this book are taken from Our World In Data at ourworldindata.org. However, the Our World In Data data set for case numbers does not extend back to the first days of January 2020 so these numbers have been constructed from contemporaneous reports as set out on Wikipedia's coronavirus timeline pages. https://en.wikipedia.org/wiki/Timeline_of_the_COVID-19_pandemic_in_January_2020.

Chapter 3

1. For an excellent round-up of the evidence on the cost-effectiveness and value of vaccinations, see https://www.gavi.org/vaccineswork/value-vaccination.

Chapter 4

1. Figures for 13 January from https://en.wikipedia.org/wiki/Timeline_of_the_COVID-19_pandemic_in_January_2020.
2. https://cepi.net/get_involved/cfps/.

3. https://www.nihr.ac.uk/news/nihr-and-ukri-launch-20-million-funding-call-for-novel-coronavirus-research/23942.

Chapter 6

1. https://www.weforum.org/agenda/2020/03/suddenly-the-er-is-collapsing-a-doctors-stark-warning-from-italys-coronavirus-epicentre/.

Chapter 7

1. Paul Offit, *Vaccinated: One Man's Quest to Defeat the World's Deadliest Diseases* (Harper Perennial, 2008).
2. https://www.uq.edu.au/news/article/2020/12/update-uq-covid-19-vaccine.
3. https://www.hhs.gov/coronavirus/explaining-operation-warp-speed/index.html.

Chapter 8

1. https://www.bbc.co.uk/news/uk-politics-52389285.
2. https://www.newstatesman.com/politics/health/2020/07/sarah-gilbert-has-shown-value-scientists-who-understand-politics.

Chapter 9

1. https://www.thelancet.com/journals/lancet/article/PIIS0140-6736(20)32657-X/fulltext?utm_campaign=lancet&utm_content=152669760&utm_medium=social&utm_source=twitter&hss_channel=tw-27013292.
2. Offit, op. cit.
3. Meredith Wadman, *The Vaccine Race: How Scientists Used Human Cells to Combat Killer Viruses* (Penguin, 2017).
4. Ibid.
5. https://www.telegraph.co.uk/only-in-britain/edward-jenner-discovers-the-smallpox-vaccine/.
6. https://www.ncbi.nlm.nih.gov/pmc/articles/PMC4328853/ and https://pubmed.ncbi.nlm.nih.gov/20563505/.
7. Leonard B. Seeff et al., 'A Serologic Follow-up of the 1942 Epidemic of Post-vaccination Hepatitis in the United States Army,' *New England*

Journal of Medicine 316 (1987): 965–70.

8. Neal Nathanson and Alexander Langmuir, 'The Cutter Incident: Poliomyelitis Following Formaldehyde-Inactivated Poliovirus Vaccination in the United States During the Spring of 1955 II: The Relationship of Poliomyelitis to Cutter Vaccine', *American Journal of Hygiene* 78 (1963): 39; Paul Offit, 'The Cutter Incident, 50 Years Later', *New England Journal of Medicine* 352 (2005): 1411.
9. https://www.sciencemuseum.org.uk/objects-and-stories/medicine/thalidomide.
10. https://understandinguncertainty.org/node/243 and http://www.numberwatch.co.uk/risks_of_travel.htm.
11. The Vaccine Knowledge Project website has more information about this: https://vk.ovg.ox.ac.uk/vk/vaccine-ingredients#Thiomersal.
12. https://britishima.org/covid19-vaccine-az/.
13. 'Ethnicity-specific factors influencing childhood immunisation decisions among Black and Asian Minority Ethnic groups in the UK: a systematic review of qualitative research', https://www.ncbi.nlm.nih.gov/pmc/articles/PMC5484038/.
14. https://journals.sagepub.com/doi/full/10.1177/1363459320925880.
15. https://www.who.int/news-room/spotlight/let-s-flatten-the-infodemic-curve.
16. http://vk.ovg.ox.ac.uk.
17. https://www.immunize.org/talking-about-vaccines/vaticandocument.htm.
18. https://www.reuters.com/article/us-health-coronavirus-trump-timeline-idUSKBN26U299.
19. Ibid., and https://www.npr.org/sections/latest-updates-trump-covid-19-results/2020/10/03/919898777/timeline-what-we-know-of-president-trumps-covid-19-diagnosis.
20. DOI:https://doi.org/10.1016/S0140-6736(20)31022-9.
21. https://www.npr.org/sections/latest-updates-trump-covid-19-results/2020/10/03/919898777/timeline-what-we-know-of-president-trumps-covid-19-diagnosis.
22. Ibid.
23. https://www.medrxiv.org/content/10.1101/2020.06.22.20137273v1.
24. https://www.fda.gov/news-events/press-announcements/coronavirus-covid-19-update-fda-authorizes-monoclonal-antibodies-treatment-covid-19.
25. https://www.bbc.co.uk/news/world-europe-55409693.

Chapter 10

1. https://www.itv.com/news/2020-07-15/positive-news-is-coming-on-oxford-covid-19-vaccine-writes-robert-peston.
2. *Telegraph*: 'Oxford scientists discover vaccine offers "double defence" against Covid', https://www.telegraph.co.uk/news/2020/07/15/coronavirus-vaccine-breakthrough-oxford-scientists-discover/. *The Times:* 'Success of early trials lifts hope for vaccine', https://www.thetimes.co.uk/article/coronavirus-vaccine-hopes-raised-by-success-of-early-trials-c2gv2cpsd. *The i:* 'All over the shop: muddle on face masks', https://twitter.com/theipaper/status/1283510998218006528.
3. https://twitter.com/NDMOxford/status/1284038977159344128?s=20.
4. https://www.express.co.uk/news/uk/1263980/coronavirus-vaccine-oxford-university-uk-covid19-testing-cure-spt.
5. https://www.ft.com/content/b053f55b-2a8b-436c-8154-0e93dcdb3c1a.
6. https://www.thetimes.co.uk/article/russians-spread-fake-news-over-oxford-coronavirus-vaccine-2nzpk8vrq.
7. https://edition.independent.co.uk/editions/uk.co.independent.issue.271120/data/9726097/index.html; https://www.telegraph.co.uk/news/2020/11/26/astrazeneca-running-new-coronavirus-vaccine-trial/; https://www.theguardian.com/world/2020/nov/26/scrutiny-grows-over-oxford-universityastrazeneca-vaccine; https://www.thetimes.co.uk/article/astrazeneca-defends-oxford-coronavirus-vaccine-as-disquiet-mounts-over-the-results-mf6t57rnr; https://www.politico.eu/article/questions-grow-over-astrazeneca-coronavirus-vax-trials/; https://www.walesonline.co.uk/news/uk-news/chris-whitty-oxford-vaccine-error-19354045.
8. https://www.handelsblatt.com/politik/deutschland/pandemie-bekaempfung-corona-impfstoff-diskussion-um-wirksamkeit-von-astrazeneca-vakzin-bei-senioren/26849788.html and https://www.reuters.com/article/health-coronavirus-eu-astrazeneca/germany-fears-astrazeneca-vaccine-wont-get-eu-approval-for-those-over-65-bild-idUSL8N2K05OP.
9. https://papers.ssrn.com/sol3/papers.cfm?abstract_id=3789264 and https://publichealthscotland.scot/news/2021/february/vaccine-linked-to-reduction-in-risk-of-covid-19-admissions-to-hospitals/.
10. See for example https://www.theguardian.com/world/2021/jan/26/

german-government-challenges-astrazeneca-covid-vaccine-efficacy-reports.

11. See for example https://www.theguardian.com/society/2021/apr/03/uptake-of-covid-jab-remains-high-in-uk-despite-blood-clot-fears?CMP=Share_iOSApp_Other.
12. *Daily Mail*, 31 December 2020, print version.
13. https://www.ft.com/content/94670990-a638-4981-84d5-283185d433b7.
14. https://www.stemwomen.co.uk/blog/2021/01/women-in-stem-percentages-of-women-in-stem-statistics and https://unesdoc.unesco.org/ark:/48223/pf0000253479.
15. https://www.vogue.co.uk/news/article/the-vogue-25-the-women-shaping-2020.
16. https://www.harpersbazaar.com/uk/culture/culture-news/g34584026/women-of-the-year-2020-winners/.

Chapter 11

1. https://www.bbc.co.uk/news/world-latin-america-55642648.
2. https://www.hhs.gov/coronavirus/explaining-operation-warp-speed/index.html.
3. https://www.nytimes.com/2020/04/22/us/politics/rick-bright-trump-hydroxychloroquine-coronavirus.html and https://www.hhs.gov/about/news/2020/05/15/trump-administration-announces-framework-and-leadership-for-operation-warp-speed.html.
4. https://abcnews.go.com/Health/timeline-tracking-trump-alongside-scientific-developments-hydroxychloroquine/story?id=72170553.
5. https://www.cnbc.com/2020/09/23/trumps-coronavirus-vaccine-czar-says-enough-after-sen-warren-says-he-should-be-fired-for-conf.html.
6. https://www.ft.com/content/b053f55b-2a8b-436c-8154-0e93dcdb3c1a.
7. https://www.ema.europa.eu/en/documents/scientific-guideline/international-conference-harmonisation-technical-requirements-registration-pharmaceuticals-human-use_en-15.pdf
8. https://www.fda.gov/media/144434/download.
9. https://www.gponline.com/malignant-tumour-caused-hpv-jab-girls-death/infections-and-infestations/infections-and-infestations/article/942531.
10. https://www.theguardian.com/world/2020/sep/09/the-oxford-university-astrazeneca-covid-19-vaccine-trial-has-been-paused-should-we-be-worried; https://www.thesun.co.uk/news/12619796/

coronavirus-vaccine-major-trial-on-hold/; https://www.reuters.com/article/us-health-coronavirus-astrazeneca-idUSKBN26017L.
11. https://www.contagionlive.com/view/sarscov2-vaccine-developers-sign-safety-pledge.
12. https://www.fda.gov/media/142749/download page 10, sub-paragraph c, and https://www.fda.gov/news-events/fda-brief/fda-brief-fda-issues-guidance-emergency-use-authorization-covid-19-vaccines.
13. https://www.theguardian.com/society/2020/oct/09/us-wont-rely-on-uk-for-covid-vaccine-safety-tests-says-nancy-pelosi.

Chapter 12

1. https://www.dailymail.co.uk/news/article-9075181/Britain-faces-calls-approve-Oxford-Universitys-coronavirus-vaccine-soon-possible.html.

Chapter 13

1. See for example https://www.theguardian.com/environment/2020/aug/05/deadly-diseases-from-wildlife-thrive-when-nature-is-destroyed-study-finds.
2. https://www.ecohealthalliance.org/2018/03/disease-x.
3. https://promedmail.org.
4. https://www.nature.com/articles/d41586-020-03518-4.
5. I lifted this line from an article by Bill and Melinda Gates. I might not have their phone number but I do read the very sensible things they have to say about vaccinations: https://www.gatesfoundation.org/ideas/articles/coronavirus-vaccine-strategy-bill-gates.

Index

adenoviral-vectored vaccines 38, 42, 48, 53, 60–1, 98–9, 113, 134–5, 154, 170–1, 212, 268, 269, 282–3
adenoviruses 30, 38–40, 52–3, 68, 101, 103, 104, 107, 108, 212, 269
adjuvant ingredient 301
Advent 87, 90, 91, 94, 100, 106, 127, 130–5, 139, 141, 143, 147, 151, 181
adverse events 165–6, 167, 174, 182, 186–7, 204–5, 244–5, 247, 253, 261
Alharbi, Naif 35
amino acids 69–70
antibodies 32, 35, 40, 60, 156, 173, 178–9, 180, 181, 186–7, 211, 214, 215, 219, 223, 239, 255–6, 260, 178, 279
Antiviral Wipe 232–3
AstraZeneca
  and efficacy of trials 12
  agreement to start manufacturing vaccine 125–30, 137–8, 145–6
  agrees to supply Covax 146
  and work on variants 279, 280–1
Atkins, Ruth 42
avian flu 30, 155, 282

B cells 59–60, 178
Baleanu, Ioana 119–20
Barrett, Amy Coney 213
Berrie, Eleanor 103, 104, 107
Biden, Joe 259
BioIndustry Association (BIA) 91, 142–3, 285
Biomedical Advanced Research and

Development Authority (BARDA) 241–3
BioNTech vaccine 203, 259
Bolam, Emma 107
Brazil/Gamma variant 21, 24, 264, 271, 278, 280
Bright, Rick 242
British Islamic Medical Association 201
Brooker, Charlie 232

Centre for Clinical Vaccinology and Tropical Medicine (CCVTM) 164
ChAd3 EBOZ vaccine 38, 40–8
ChAdOx1 vaccines 48–9, 50, 51, 54, 55–6, 64, 65, 73–4, 91, 100–1, 129, 146, 154, 170–1, 175, 178, 186–7, 202–3, 205, 238, 253
clarified cell lysate 113–14
Clinical Biomanufacturing Facility (CBF) 42
  CG's work at 58, 60–1
  work on vaccines 59, 60–1
  design phase of vaccine 62–3
  staring materials for vaccine 72–3, 74, 100
  in process for making vaccine 100, 106, 107, 108, 110, 113, 119, 120, 127, 130, 133–4, 140–1
  and clinical trials 138, 139, 159
  during lockdown 185
  visit from Prince William 189–92
clinical-grade vaccines 58, 62, 73, 86, 106, 159
clinical trials 9, 18, 20, 21, 25, 27, 41, 54–5, 78, 81–2, 85, 86, 87–8, 90, 94, 100, 112–13, 129–30, 133, 134, 135–6, 141–2, 150, 153–6, 157, 158–61, 163–87, 223, 240–4, 250–1, 252–3, 265, 271–2, 279
Coalition for Epidemic Preparedness and Innovation (CEPI)
  SG's work for 50–4
  funding from 82, 83, 85–6, 89, 91, 92, 94, 156
  and Covax 146
codons 69–70
conjugate vaccines 301–2
Cornall, Richard 173, 175
coronavirus
  description of 68

Covax 94, 146, 241, 287–8
Covid–19
  early reports on 29–30, 55, 61–2
  transmission of 35–6, 71, 281–2
  lockdown in UK 77, 110–11, 112, 262–3
  declared public health emergency by WHO 85–6
  declared global pandemic by WHO 93
Crimean-Congo haemorrhagic fever 48, 282
Cunk, Philomena 232–3
Cutter Laboratories 195

Daily Mail 230, 263
Department of Health (DoH) 94
design phase in vaccine development 62–7, 102
dexamethasone 214–15
Di Marco, Stefania 131
Diamond Princess 110
Disease X 51–6, 61–2, 153
DNA
  in design of vaccine 65, 67–72, 73–4, 85, 154
  in manufacture of vaccine 100–1, 103, 104–5, 265, 275–6
  and vaccine hesitancy 192, 203
DNA vaccines 53, 85, 86, 195, 302
Doremalen, Neeltje van 55
Douglas, Sandy 90, 106, 115, 126, 140–2, 144–5, 231

Ebola virus
  2014 outbreak of 36–7, 84
  vaccine development for 38, 39–47, 61, 86–7, 107–8, 155, 282
  method of spread 38–9
  2021 outbreak of 47
  2020 outbreak of 282
efficacy of trials 11–12, 15–18, 22, 25, 26, 142, 152, 160, 172, 181, 225–6, 227, 239, 241, 249, 250, 260–2, 279
El-Muhanna, Omar 137
Elizabeth II, Queen 21
emergency–use licensure 19, 152, 160, 250–1, 259–61, 263, 265–7, 270
emerging pathogens 30–1
EU vaccination programme 22–3, 26
European Medicines Agency (EMA) 21, 22–3

Fauci, Anthony 15
Financial Times 244
Foderato, Denise 179
Food and Drug Administration (FDA) 215, 238, 243, 249, 250–1, 252–3
funding
  for vaccine development 77–95, 141–3, 144–5, 156–9
  for future pandemics 283–4

Gates Foundation 156
Gavi 146, 156
Gilbert, Sarah
  and clinical trials 6–7, 8–15, 164, 175, 183
  and early media reaction 16–17
  and effectiveness of vaccine 24–5
  early awareness of Covid–19 29–30, 55
  research on pathogens 30–1
  development of vaccine for Ebola 36–7, 41–2
  development of ChAdOx1 vaccine 48–9
  work with CEPI 50–4
  work on MERS-CoV vaccine 51, 172
  work on Disease X 51–6, 61, 153
  work on design phase of vaccine 62, 63–4, 65, 102
  and Sarah Sebastian 68
  work on staring materials for vaccine 71–5, 98, 102
  securing funding for vaccine 77–95, 141–2
  treated for bile acid malabsorption 79–80, 88–9
  appears at Science and Technology Committee 92–3
  in process for making vaccine 98, 102, 106, 151–5, 158
  work/life stresses of 149–52, 256–8
  media appearances 176
  and visit from Prince William 189, 190, 191–2
  on vaccine hesitancy 192–217
  and media reaction to clinical trials 223–4
  media interest in 231, 232
  and results of clinical trials 237–8, 242
  work on variants 264–5, 271

and licence for vaccine 265–7
has first vaccination 271–3
return to other projects 288
Ginsburg, Ruth Bader 213
good manufacturing practice (GMP) 58, 103, 143, 144
Granato, Elisa 164, 176, 222–3
Green, Catherine
encounters vaccine hesitancy 1–4
university work 58, 60–1
early awareness of Covid–19 61–2
work on staring materials for vaccine 71–5, 86–7, 89–90, 91
trip to Paris 75
and funding for vaccine 77
on process of making vaccine 98–123, 130–5, 140–3
agreement with AstraZeneca 125–30
and clinical trials 138, 164, 168, 173–4, 176–7, 182, 183
media appearances 163, 176
effect of lockdown on 184–5
and visit from Prince William 191
and media reaction to clinical trials 219, 220, 221–2, 226
on *Panorama* programme 228–30
media interest in 230, 231
work on variants 265, 271
return to other projects 288
Guardian, The 247–8

Harper's Bazaar 232
HEK293 cells 101, 109, 111, 113, 132, 212, 215–16, 268
Hill, Adrian 106, 145, 231, 252
Hilleman, Maurice 153, 154–5, 193
HLA molecules 268–70

immune response of vaccine 20, 70, 86, 113, 134–5, 154, 166–7, 170–1, 172, 174, 175–6, 178, 180, 181, 182, 183–4, 185–7, 208, 227, 239–40, 243, 251, 253, 254, 260, 264, 271, 279
inactivated vaccines 300

Janssen (company) 49
Janssen vaccine 23, 26–7, 135

Jenner, Edward 60, 194, 221
Jenner Institute 41, 42, 152
Joe, Carina 141
Johnson, Boris
  and January 2021 lockdown 20
  announces route out of lockdown 26
  in self-isolation 77
  fails to attend COBRA meetings 104
Johnson & Johnson *see* Janssen vaccine
Joint Committee on Vaccination and Immunisation (JCVI) 20–1

Kent/Alpha variant 19, 24, 262, 264, 271, 277–8

Lambe, Teresa (Tess)
  and efficacy of trials 13
  and genetic sequence of Covid-19 55, 67–8, 70
  work on Ebola vaccine 61, 86–7, 282
  work on design phase of vaccine 62, 63–4
  work on staring materials for vaccine 71–2, 78, 84
  in process for making vaccine 106
  media interest in 231
Lancet 17–18
Lassa fever 48, 50, 54, 72, 88, 91, 92, 155, 282, 283
  emergency–use licensure 19, 152, 160, 250–1, 259–61, 263, 265–7, 270
live attenuated vaccines 299–300

Macron, Emmanuel 22
Marburg virus 48, 54, 72
media reaction to clinical trials 15–18, 219–28, 247–8
Medicines and Healthcare products Regulatory Agency (MHRA)
  emergency–use licensure 19, 152, 160, 250–1, 259–61, 263, 265–7, 270
  role of 20
  and process for making vaccines 108–9, 121, 122
  and clinical trials for vaccine 135–6, 238
  and safety of Pfizer vaccine 203–4
MenACWY vaccine 175
MERS-CoV
  spread of 31–4
  in South Korea 34

as continuing issue 34–5
vaccine development for 50, 51, 64–5, 171, 172
transmission of 281–2
2020 outbreaks 282
methionine 69–70
mixing vaccines 20–1
Moderna vaccine
efficacy of 11, 15, 259
and early media reaction 18
and EU vaccination programme 22
roll–out in United States 26–7
clinical trials of 245
Morris, Sue 74, 100, 102
mRNA vaccines 53, 85, 195, 203, 243, 283
Munster, Vincent 55

next-generation sequencing (NGS) 104
Nipah virus 48, 50, 54, 88, 91, 92, 155, 282, 283
Novavax vaccine 23
Nuffield Department of Medicine 61

O'Brien, James 163
Official Development Assistance 84, 284
Oliveira, Cathy 107, 108, 109, 116, 117
O'Neill, Edward 164
Operation Warp Speed (OWS) 158, 241–2, 243
Oxford AstraZeneca vaccine *see also* vaccines for Covid–19
clinical trials 6–7, 8–15, 26, 28, 159–60, 163–87, 219–28, 235–48
media reaction to 15–18, 219–28, 247–8
emergency–use licensure 19, 152, 160, 250–1, 259–61, 263, 265–7, 270
roll–out in UK starts 20–1
and EU vaccination programme 22–3, 26
roll–out in UK speeds up 23
distribution around world 24
effectiveness of 24–5
roll–out in United States 26–7
agreement to start manufacturing 125–30, 137–8
large–scale manufacturing starts 144–7
and Operation Warp Speed 158, 241–2, 243

ingredients of 201
in Germany 227–8
efficacy with variants 264–5, 279
issue with HLA molecules 268–70
description of 311–14
Oxford BioMedica 145, 146
Oxford Vaccine Group 87, 165

Pall facility (Portsmouth) 143, 144
Panorama programme 228–30
Parracho, Helena 120
pathogens
description of 30–1
PCR test 255, 256
Pelosi, Nancy 250
Peston, Robert 219, 224
Pfizer vaccine
efficacy of 11, 15, 259
and early media reaction 18
roll-out started 18–19
and EU vaccination programme 22
effectiveness of 24–5
roll–out in United States 26–7
safety of 203–4
phase I trials 42, 48, 135–6, 138, 140, 156, 159, 165–8, 170, 175–6, 179–80, 185–6, 239, 240, 272, 279
phase II trials 42, 46, 133, 135–6, 137, 140, 166–7, 180, 181–2, 239
phase III trials 17–18, 42, 44–5, 46, 85, 94, 157, 160, 167–8, 180–4, 205, 208, 225–6, 239–40, 243–4, 250
Philip, Duke of Edinburgh 21
Phipps, James 194
placebo in trials 164, 167–8, 175, 211, 238, 240
platform technologies 48–9, 52–3, 54, 60–1, 64, 85, 154–5, 170–1, 195, 278–9, 282–3, 284, 302–4
Pollard, Andy 231
and clinical trials 9–10, 11, 87–8, 165, 246, 271
and vaccine roll–out 20
in process for making vaccine 106
and agreement with AstraZeneca 125–6
and visit from Prince William 189–90, 191
media appearances 176, 232–3

recovering from work on vaccine 288–9
Popova, Olga 49, 50
preclinical trials 85, 159, 160, 173, 208
production cells 109, 110, 112, 113, 132, 133, 145, 268, 279
Public Health England 1–2

Quattrone, Frank 179
Queensland, University of 156

recombinant protein vaccines 300–1
RECOVERY trial 214
replication-deficient recombinant simian adenoviral-vectored vaccine 38–9, 45–6, 48, 49, 64, 170–1, 202–3
replication-deficient viruses 38–9, 40, 48, 49, 195
research-grade vaccines 62–3, 143, 159
Rift Valley fever 48, 54
Rocky Mountain Laboratories 55, 175

SARS
similarity to Covid-19 30
spread of 31
eradication of 34
as WHO priority disease 48
development of vaccine for 174
transmission of 281–2
SARS-CoV-2
comparison to Ebola 37
and design of vaccine 64–5, 67
work on genome sequence 62, 67–8, 70
manufacture of vaccine 154
and trials for vaccine 172, 173, 178, 186, 204, 224, 239, 255, 259
and vaccine hesitancy 202–3, 204, 211, 289
worldwide effect of 206
and Moderna vaccine 243
variants of 264–5, 276–81
difficulty of containing 283
Science Media Centre 225
Sebastian, Sarah 68, 70, 73, 73
Serum Institute India (SII) 21, 145
single-virion cloning 102–3
SK Bio 146
Slaoui, Moncef 242–3

South Africa/Beta variant 21, 24, 271, 277, 278, 280
Spencer, Alex 134–5
spike proteins 64–5, 67–8, 70, 73, 74, 100, 101, 138–9, 154, 173, 202, 205, 211, 214, 224, 239, 248, 264, 277–8, 280
starting materials 65–7, 71–5, 86–7, 91, 98, 100–5, 132
STIKO 21
subunit vaccines 300–1
'Summary of Product Characteristics' sheet 201, 209
Sun, The 248
suspected unexpected serious adverse reaction (SUSAR) 244–8

Tarrant, Richard 123
T cells 59, 178–9, 180, 186–7, 219, 260, 279, 280
toxoid vaccines 301
transfections 101–2, 104
trials for vaccine
clinical 9, 18, 20, 21, 25, 27, 41, 54–5, 78, 81–2, 85, 86, 87–8, 90, 94, 100, 112–13, 129–30, 133, 134, 135–6, 141–2, 150, 153–6, 157, 158–61, 163–87, 223, 240–4, 250–1, 252–3, 265, 271–2, 279
efficacy of 11–12, 15–18, 22, 25, 26, 142, 152, 160, 172, 181, 225–6, 227, 239, 241, 249, 250, 260–2, 279
phase III 17–18, 42, 44–5, 46, 85, 94, 157, 160, 167–8, 180–4, 205, 208, 225–6, 239–40, 243–4, 250
phase II 42, 46, 133, 135–6, 137, 140, 166–7, 180, 181–2, 239
phase I 42, 48, 135–6, 138, 140, 156, 159, 165–8, 170, 175–6, 179–80, 185–6, 239, 240, 272, 279
preclinical trials 85, 159, 160, 173, 208
placebo in 164, 167–8, 175, 211, 238, 240
Trump, Donald 27, 213–15, 224, 241, 243, 244, 253
Turner, Jonathan 179
Turner, Tracy 179
Tuskegee experiment 193–4

UK Genome Stability Network 61

UK Research and Innovation (UKRI) 78, 82, 85–6, 89, 90, 91, 94, 141–2
United States of America
  vaccine roll–out in 26–7
  and Operation Warp Speed 158. 241–2, 243
  and Tuskegee experiment 193–4
  clinical trials in 240–4, 250–1, 252–3
  and America First stance 241

vaccine hesitancy
  CG encounters 1–4
  history of 193–5
  fear of contamination 195–6
  understanding of risk 196–200
  fears of vaccine ingredients 200–2
  safety of vaccines 202–4
  and pregnancy 204
  within ethnic minority communities 204
  possible side-effects of vaccine 204–8
  fear of blood clots 205–8
  and later exposure to pathogen 208
  trust in professionals 208–9
  role of social media in 209–12
  use of foetal cells in vaccines 212–13
  in Germany 227–8
Vaccine Knowledge Project 211–12
vaccines
  clinical-grade 58, 62, 73, 86, 106, 159
  effectiveness of 58–9, 182
  different ways of working 60–1
  research-grade vaccines 62–3, 143, 159
  process for making 98–123, 305–10
  development time for 152–5
  trial process for 165–8
  in preparation for next pandemic 281–6, 288–9
  types of 299–302
vaccines for Covid–19 *see also* Oxford AstraZeneca vaccine
  design phase 62–7, 102
  starting materials for 65–7, 71–5, 86–7, 91, 98, 100–5, 132
  adding genome sequence to 68–70

funding for 77–95, 141–3, 144–5, 156–9
process for making 98–123
initial manufacture of 106–12, 127, 130–5, 140–3, 151–5
purifying vaccine 112–18, 132–3, 141–2
put in sterilised vials 118–20
vial labelling 120–1, 133
documentation of process 121
quality certification 122–37
speed of development 153–61
Vaccines Manufacturing and Innovation Centre (VMIC) 126, 143, 145, 285
Vaccines Taskforce 94, 111, 144–5
Vaccitech 68
variants of SARS-CoV-2 264–5, 276–81
VaxHub 80, 84–5
Vegebroth 63
Velicka, Oto 183
virus-like particle vaccines 300–1
virus particles 101–4
viruses
description of 59–60
Vogue 232
VSV vaccine 40, 45, 47, 87

Walsh, Fergus 164, 176, 228, 229
Wellcome Centre for Human Genetics 58, 61, 138
Wellcome Trust 82–3
Whitty, Chris 164
William, Prince 13–14, 189–92
World Health Organization (WHO) 61
approves vaccine for use 24
draws up list of dangerous pathogens 48–9, 51
declares Covid-19 public health emergency 85–6
declares Covid-19 global pandemic 93
and timetable for vaccine 106
and Covax 146
United States withdraws from 241
early-warning system 286